Basic Concepts in Fruit Science

NIPA® GENX ELECTRONIC RESOURCES & SOLUTIONS P. LTD.
New Delhi - 110 034

About the Authors

Dr. Niranjan Singh (Ph.D.) is currently working as an Assistant Professor in the Department of Fruit Science, Acharya Narendra Deva University of Agriculture and Technology, Ayodhya (UP), India. Born on 15 July 1988 in Gyanagarh, a small village in Saharanpur, UP, he obtained Bachelor's degree in Agriculture from CCS University, Meerut, UP in 2009, M.Sc. (Ag) in Horticulture from CSA University of Agriculture and Technology, Kanpur, UP in 2012 and doctorate in Fruit Science from Dr. YS Parmar University of Horticulture and Forestry, Solan, HP in 2017. He worked as SRF and RA under World Bank funded Project of Himachal Pradesh Horticulture Development Project operational in the Department of Fruit Science, Dr. YSPUHF (HP) from 2018 to 2022 and as a Subject Matter Specialist (Horticulture) in Krishi Vigyan Kendra, Moradabad-II, UP under the University of SVP University of Agriculture and Technology, Meerut, UP in 2022. He has qualified ICAR-JRF, SRF, and NET. A receipient of *Young Scientist Award* given by Genesis Urban and Rural Development Society (Hyderabad, India), he has published 24 research papers in national and international journals, 8 popular articles, 4 review papers, 2 practical manuals, 1 lead paper, 1 book, and 14 book chapters.

Dr. D.P. Sharma, born on 1st June, 1967, did his B.Sc. (Agri.) from HPKV, Palampur, College of Agriculture, Solan in 1988, MSc. and Ph.D. in the discipline of Pomology (Fruit Science) from Dr. YS Parmar, University of Horticulture and Forestry, Nauni, Solan in 1991 and 1995, respectively. Started his carrier as Assistant Scientist (Horticulture) at RHRSS-Tabo in Lahul & Spiti, Himachal Pradesh and served tribal area for 12½ years. During his tenure at Tabo, research work was commended by the Board of Management of University in its 57th and 67th meetings for his excellent work. Served as Principal Scientist and Head at HRTS and KVK, Solan at Kandaghat and grabbed first KVK award for the year 2019. Presently as Professor and Head in the Department of Fruit Science (Dr. YSP UHF Nauni Solan HP-173230). Guided 6 Ph.D. and 18 M.Sc. students in the discipline of Fruit Science. Managed more than Rs. 562 lakh funds through 16 adhoc research project as Principal Investigator from outside funding agencies. Pioneer in installation and demonstration of solar power fencing model in Department of Fruit Science, which was adopted by State Government of Himachal Pradesh in the year of 2015-16 as Mukhya Mantri Phasal Surksha Yojna. Visited USA during the years 2021, 2022 and 2023 as member of State Level Pre-dispatch Inspection Committee (PDI) for import of planting material of temperate fruits under Himachal Pradesh- Horticulture Development Project World Bank funded. His field of specialization is Orchard Management and has more than 28 years experience in the scientific management of orchards specially fruit nutrition, canopy management and nursery growing. Introduced lateral bearing varieties of walnut presently working on replant problem in apple and peach as well as high density planting of apple. He has co-authored a book entitled "Systematics of fruit and applied fruit breeding techniques", one as author *E-book* on "Plant propagation and nursery management", 40 book chapters and 77 research and review papers to his credit published in National and International Journals.

Neeraj Sankhyan, PhD, is currently working as an Assistant Professor at Dr YSPUHF, Solan, (HP) where he teaches English Language, Communication Skills and Technical Writing. He graduated from Delhi University in arts, earned his master's from IGNOU and doctorate from IIT Mandi (HP) in English Literature. His doctoral research was based on the thematic study of English poets from the conflict-affected areas of the Indian Himalayas. He is passionate about teaching, language and literature.

Basic Concepts in Fruit Science

Niranjan Singh
Assistant Professor
Department of Fruit Science
Acharya Narendra Deva University of Agriculture and Technologya
Ayodhya (UP), India

D.P. Sharma
Professor and Head
Department of Fruit Science
Dr. Yashwant Singh Parmar University of Horticulture & Forestry
Nauni, Solan, Himachal Pradesh

Neeraj Sankhyan
Assistant Professor
Dr Yashwant Singh Parmar University of Horticulture & Forestry
Nauni, Solan, Himachal Pradesh

NIPA® GENX ELECTRONIC RESOURCES & SOLUTIONS P. LTD.
New Delhi - 110 034

NIPA® GENX ELECTRONIC
RESOURCES & SOLUTIONS P. LTD.

101, 103, Vikas Surya Plaza, CU Block
L.S.C.Market, Pitam Pura, New Delhi-110 034
Ph : +91 11 27341616, 27341717, 27341718
E-mail:newindiapublishingagency@gmail.com
www: www.nipabooks.com

For customer assistance, please contact
Phone: + 91-11-27 34 17 17 Fax: + 91-11- 27 34 16 16
E-Mail: feedbacks@nipabooks.com

ISBN: 978-81-94766-86-5

Composed and Designed by NIPA.

Preface

Pomological crops include tropical, sub-tropical, temperate fruits, minor fruits and plantation crops. The present book **"Basic Concepts in Fruit Science"** is an up-to-date, enlarged comprehensive, and advanced book on these crops with an overview of advances in crop production and improvement; classification, nomenclature, systematics, genetics, breeding, crop improvement, and crop production management of fruit crops. All this information is presented point-to-point for easy understanding by the students, researchers, and teachers. This book is designed to serve as a handy guide for students, teachers, researchers, and amateurs engaged in Fruit Science. It will serve as a guide to the students and teachers to test their knowledge in the field of fruit science as a whole.

We believe that this book will also be useful for fellow students, teachers, researchers, and development officers for reference and easy answering of many complicated questions. Special emphasis has been given to students appearing for various competitive examinations like post-graduate, SAUs entrance, Ph.D. Fruit Science, Junior Research Fellowship (JRF), Senior Research Fellowship (SRF), Agricultural Research Service (ARS), ICAR-National Eligibility Test (NET) Fruit Science, Horticulture Development Officer, Horticulture Extension Officer, Civil Services, Agricultural Officers, Bank P.O. (Agriculture), Agriculture Development Officer and Allied Agricultural Examination Services.

We are grateful to all those persons as well as books, manuals, periodicals, magazines, newsletters, journals, etc. that helped in the preparation of this book. Despite the best efforts, some errors may have occurred in the compilation and editing of the book. Further queries, constructive suggestions, and criticisms for improvement of the book are always welcome and shall be thankfully acknowledged. Last, but not least, it is a pleasure for us to extend our sincere thanks to **NIPA Publications** and team members for their keen interest shown in the preparation and publishing of this book so efficiently and promptly.

Authors

Contents

1

Introduction and Research of Fruit Crops

- Fruit research in India was started in the department of Botany in six-agriculture College in 1905.
- A pomological station was established at Coonoor, Ooty in 1920 to study the adaptability of temperate fruit varieties.
- Four Regional Research Station in – 1933

Subtropical location at

- Sabour, Bihar
- Koduru, Madras
- Krishnanagar, Bengal
- Temperate Region at Chaubatia, Uttarakhand
- The research was also initiated in 1945 in Jammu and Kashmir by the *Survey of deciduous fruits*
- In 1950, Horticulture Research Station was established at Kandaghat and Kullu
- 1956 in eight research stations were started:
    - Saharanpur, Uttar Pradesh
    - Mashobra, Himachal Pradesh
    - Abohar, Punjab
    - Kalakuchi, Assam
    - Pune Maharashtra
    - Kodur, Andhra Pradesh
    - Chettalli, Mysore, Karnataka
    - Sabour, Bihar
- Horticulture Research Station, Tamil Nadu – 1961.

- Division of Horticulture in IARI in 1956 and separate Division of Fruits in 1984.
- Research Virus disease of fruit cross at Pune was established in 1958 to
- Shimla to conduct research or rootstocks of stone and pone fruits.
- IARI Regional Station, Pusa (Bihar) carried out Research on Papaya and Litchi.
- Indian Institute of Horticulture Research Bangalore 1968.
- Central Horticultural Experiment Station, Chettalli situated in Kodagu district of Karnataka established in 1947 was transferred to the IIHR on February 11, 1972.
- Central Mango Research Station – 1972.
- March 1979 to work on the problems of tribal area of western India. The CHES Ranchi was established in May 1979 to work on the problems of the tribal areas of eastern India.
- CIHNP was again renamed as CISH 1995.
- National Bureau of Plant Genetic Resources – 1978. (NBPER-1978)
- All India Coordinated Fruit Improvement Project stated in 1971
- Headquarter of (AICFIP) in CMRS Lucknow in 1972.
- Reorganized as (AICRP) on subtropical and temperature fruits in 1986.
- Institute of Temperate Horticulture at Srinagar in 1993.
- During the Eighth plan, a center has been proposed to be started on postharvest technology of Arid Zone fruits in Rajasthan.
- The ICAR sanctioned two centers of advanced studies in 1979.
- Centre of advanced studies in Tropical Horticulture at Bangalore functioned unit 1986.
- Tissue culture methods for clonal multiplication in Papaya, Banana, Date etc.
- Oldest fruit: Date Palm.
- New young fruit: Avocado
- Oldest fruit Literature: Litchi.
- The first book in the world on fruit culture – Litchi.
- Industrialization on a large scale is a product of the 19th and 20th centuries.

- Aurangzeb was fond of mangoes and named two of his favourite varieties Sidhras and Rasna Vilas.
- Pome meaning – fruit
- Ology meaning – a branch of learning
- Father of systematic pomology – De Candole
- 143 men days/ha/annum for cereal crops but in horticulture 860 man-days/ha/annum are required
- Maximum man-days are required in Grape, Banana and Pineapple upto 1000- 2500 Monday/ha/annum, greater much higher employment
- India has attained the first position in 1994 among coconut production
- Horticulture products account for over 25% of the total export of agricultural commodities from India.
- Export of Cashew lands the horticulture commodities.
- Export of cashew kernels roughly 50% of total demands.
- Only about 30-35% of the price paid by the consumers goes to the grower.
- The cooperatives can handle only about 4-5% of total production.
- National Agriculture Cooperative Marketing Federation of India (NAFED) at New Delhi.
- Date palm is the most salt-tolerant it can tolerate 4% salt and can withstand as much a 12000ppm chlorides in irrigation water.
- Sikkim produces most of the large cardamom over 90% of the total production of the country.
- Apple can be kept for 90-240 days in variable temperature
- The shelf life of strawberries even at a freezing point is only - 5-7 days.
- Among the micronutrients deficiency of Vit.-A. Iron and Iodine are recognized as public health problems.
- Essential plant nutrients:- Carbon, Hydrogen and Oxygen.
- Mineral nutrient – other 14 nutrients.
- Phosphorous: Important to root growth and seed germinating and deficiency more frequent in acid soils.
- Potash: Especially important for fruit crops, tomatoes, potatoes and flowers. Deficiency is more frequent in sandy soils.
- Fruit varieties for rainfed areas:

Crop	Varieties
Ber	Gola, Seb, Mundia, Banarasi, Kandaka
Aonla	Chakaiya, NA.6, NA.7, Balwant, Kanchan, Krishna. Pomegranate Ganesh, Muskat Dholka, Jalore Seedless, C-13 Jyoti Guava Sardar, Allahabad Sefeda
Custard apple	Balanager, Mammoth
Ber:	
Z. mauritiana	*Z. jujuba*
Indian Ber	Chinese Ber
Native place India	Native place China
Vegetative growth rainy season up to December	Vegetative growth start in Spring
Sheds leave in summer	Sheds leaves in winter Flowering – September to November Flowering – April-June Flower born in a cluster of leaf axil Flower bud appears or cluster Fruits mature in Jun, Feb-March Fruit mature Aug-Oct

- Ber seeds contain saponin, jujubogenin
- Ellaichi variety of Ber is almost seedless aromatic on suitable for the kitchen garden, pollen sterile, octoploid.
- Most vigorous rootstock: *Z. rotundifolia*
- Fruit setting in ber generally 2-6%.
- Ber is a single sigmoid growth curve
- Ber is also king of arid gone fruit Chinese fig. Chinese poor's man fruits.
- Litchi first time introduced in Bengal in 17th century
- The red colour in litchi is due to cyanidin (anthocyanin)
- In aonla presence of Vitamin C. (500-750mg/100g pulp)
- In aonla lower portion flowers usually female and in upper portion flowers usually male.
- Kanchan (NA-4) maximum female flower per branchlet.
- Chakaiya is very prolific in bearing.
- Krishna had a maximum viable pollen (92.4%).
- Phalsa is also known as filler tree.
- Losora is also called Indian cherry.
- Auxin called master hormone.

- Most commonly utilized hormone is IBA (typical rooting hormone) followed by IAA.
- Highest longevity of fruits plant: Chestnut seasonal (annual) cape gooseberry, 1-2 years – Pineapple, Banana.
- Red value of seed (R.V) Purity % × germination / 100
- Day neutral fruit plant – Banana, Papaya, Coconut
- Some fruit crops do not produce viable seeds e.g., banana, fig and pineapple.
- Softness in mango is due to a calcium deficiency whereas softness in banana is due to N deficiency.
- The aroma of fruits is due to
  - Banana: Green – Hexanal, Ripe – Eugenol, Overripe – Isopentenyl acetate.
  - Orange – Valencia
  - Grapefruit – nootkatone
  - Cherry, Peach, Almond: Benzaldehyde
  - Apple green – Hexanal
  - Apple Ripe – Ethyl-2-methyl butyrate
  - Lemon – citral/geraniol
  - Cranberry – 2-methyl butyrate acid
  - Grape – Methyl Salicylate
  - Strawberry – Ethyl butyrate
  - Raspberry- Butyl acetate
  - Clove – Eugenol
  - Vanilla – Thymol
  - Penicillium – Penicillin
  - Aspergillus – Ka blue mold
  - Mucor – Grey mould/in mould/ braille mould.
- Most fruits are acidic when mature (<4.5pH)
- Grape wine is the oldest example of a fermented beverage.
- Lawerkravt means arid cabbage, it is popular in Europe and USA
- Fruit chess is also called Karachi Halwa. It is made using fruits such as guava, plum, apple, pear etc.

- By tissue culture/micro propagation, 0.5 million plants were produced in 1987 using tissue culture or micropropagation. However, presently, over 25 million plants are produced using this method.
- On an average, 50 tissue culture lab/units are operating in our country.
- Karnataka has been given the status of thrust sector industry to plant tissue culture ventures.
- The first commercial micro propagation laboratory was set up in 1984.
- Synseed is a synthetic seed.
- Coconut cultivation in India uses maximum drip irrigation.
- True potato seed technology in India has so far been successful only in Tripura.
- IFOAM – International Federation of organic agriculture movement.
- Its headquarters are located at Thale in Germany
- DRDO – Field Research Laboratory Leh, installed a glasshouse during 1964. It is perhaps one of the oldest greenhouses in the country.
- BIPM – Bio-Intensive Integrated Pest Management.
- ICM – Integrated Crop Management System.
- Gene transformation can be achieved a lot more efficiently and regularly in dicots than in some monocots.
- Bt. Coat protein, Chitinase and glucanase are frequently used foreign genes for developing transformation plants which improves their tolerance to specific pathogens.
- The duration of misting should normally be around 10 seconds for every 20 minutes of mist propagation.
- At present, very little area (approx 0.2%) of cultivated land in India under the cultivation of True Potato Seed (TPS)
- TPS crop takes about 20-25 days more than those taken by tubers seeds to mature.
- 'P' series rootstocks derived from cross between EMI with Antonoka *viz*., P1, P2, P5 and P22 are used as apple rootstock in Poland.
- St. Julia is a semi-dwarfing rootstock for plum (Dwarfing rootstock for plum) was developed by EMRS Kent (England)
- Flying dragon is a good dwarfing rootstock of citrus fruits.
- 'Tatura Trellis' system was developed by Irrigation Research Institute, Tatura, Australia, in 1973. It consists of rows of 'V' shaped trees running

North to South in such a manner that each tree has only two limbs growing east and west at an angle of 60^{0} to the ground level.

- Hedgerow system is the most modern method for apple, pear and peach cultivation.
- Dikegulac is a chemical that is very effective in enhancing lateral branching and thereby reducing the size in several woody perennials.
- A single dominating gene designated as 'CO' in apple and 'DW' in peach has been superimposed on the polygenic control.
- DIRS- Diagnosis and Recommendation Integrated System was proposed by Beau Fils, 1973
- Most of the applied phosphorus creates chemical and physical precipitation which leads to clogging problems.
- It is estimated that about 27 million ha area in our country is suitable for drip irrigation.
- In India, about 90% water consumption is done for the agriculture system.
- India has the largest irrigation network in the world.
- Irrigation efficiency in our country has not been more than 40%. The gross irrigated area in the country is only about 38%.
- Papaya (40-100%) among fruits and Cauliflower (49%) amongst vegetables suffers the maximum post-harvest losses fruit and in vegetable.
- Over 80% of the 5000 cold storage in India is used for potato.
- Cleaning is the of the process line in a plant processing fruit and vegetable.
- The USA is the largest consumer of dried and artificial flowers.
- Grapes and vegetables are highly contaminated with pesticides.
- First HDP of appl e was established in Europe in the sixties.
- Colt, a semi vigorous rootstock of cherry is very popular.
- The greenhouse could well be called "Food factories".
- Aero ponies – A modified form of hydroponics is going to be the technology for future. In this technology, nutrient mist is provided periodically at plant roots.
- Coconut root (wilt) of *Phytoplasma* is the causative agent of the disease. The disease is transmitted by Lace bug, *Stephanitis typicus* and plant hopper, *Proutista moesta.*
- Trifoliate orange (*Poncirus trifoliate*) and its hybrids (citronella) are highly resistant to the citrus nematode.

- IIHR has been derived by crossing Rangpur line × *Poncirus trifoliate*) *viz.* CRH-3, CRH-5 and CRH-41 highly resistant *C. nematode* CRH- Citrus Rootstock Hybrid.
- The credit for success of the first classical biological control goes to India. It was started as early as 1762. The Indian Myna bird (*Acridotheres tristis*) was exported to Mauritius and it could successfully control the red locust *Nomadacris septemfasciata*.
- BIPM technologies promote growing of trap border crops to conserve natural enemies.
- 10-50 beetles are generally released per plant/tree in horticultural and plantation crops depending upon the crop canopy and intensity of mealy bug infestation.
- Percentage proportion of Pesticide use in India: 71% in the form of insecticide 18% herbicide 6% fungicide 5% others
- Percentage of pesticide use in the world:45% Herbicides 36% insecticides 17% Fungicides 2% others
- According to a survey conducted by ICMR, 51% of our food for commodities were contaminated with pesticides residues.
- The loss of fruit yield due to pest and diseases has been estimated to be around 13- 34%. The percentage of pesticides applied on fruit crops is very low (<10%) as compared to field crops.
- Among the fruit crops, grapes vines consume the maximum share of pesticides particularly fungicides.
- Apple scab (*Venturia inaequalis*) is the most serious problem of apple and is prevalent throughout India.
- Sorting can be defined as of the commodity into a various group of varying physical properties by size and shape.
- Grading is based on quality and is done using separation.
- Water used for hydro cooling is generally kept between 0°C and 0.5°C.
- In vacuum cooling, there is generally reduction in the weight of the product (by 1%) due to moisture loss resulting in a temperature decrease of 6°C.
- Forced air cooling can cool the produce six to ten times faster than room cooling. In airflow, air is blown over warm produce at the rate of 0.5 to 3 litres per second per kg (15-1 kg-1) for sufficient heat removal.
- Room cooling has been categorized as a pre-cooling technique.

- Flashing, a term used for a sudden drop in the temperature is caused by a sudden pre-drop.
- Underground storage is good for keeping the produce cool but it does not remove the field heat.
- The groundwater temperature usually does not vary more than 1°C during a year for any given location.
- C_2H_4 will induce ripening in many fruits and can also cause some physiological disorders in vegetables. C_2H_4 produced by the commodity can be reduced by decreasing the surrounding O_2 levels and increasing the CO_2 levels.
- TQM (Total Quality Management)
- UV light reacts with O_2 to form O_3 (Ozone) which can destroy C_2H_4 (ethylene).
- The fruit fly is a very harmful pest of ber and guava.
- At present, the total cold storage capacity in India is about 87 lakh tonnes.
- Vacuum cooling is suitable for leafy vegetables while hydro cooling is effective for tubers.
- Controlled Atmosphere (CA) is one of the most advanced techniques of storage.
- Japan is the leading country in the world to have achieved TQM in industry.
- National Institute of Research Management of ICAR is situated in Rajendranagar, Hyderabad India.
- Horticultural crops cover about 12 million ha area and contribute about 18.8% to the gross domestic product and 15.5% of the total agricultural export.
- Sprinklers and drip irrigation systems are jointly known as pressurized irrigation system.
- Approximately, 60% of the horticultural crops are consumed by the local population or marketed in the near markets and only 40% of the produce is channelized in the regulated markets for the consumption of the urban population in big cities.
- Export market accounts for approximately 1% of the production. Less than 1% of the horticultural produce is commercially processed in Indian and international markets.
- Transgenic soybean ranks first globally in terms of area under cultivation. It occupies 52% of the total area under transgenics followed by corn (30%), cotton (9%), Canola (9%) and potato.

- Gene transfer to fungi is generally much more difficult than gene transfer to bacteria.
- India's share in the total world production of fruits is 12%.
- Word floriculture trade is still controlled by flatland which has the lion's share of 67% to its credit.

Fruits	Vegetables
1. Fruit plants are perennial	Most vegetable plants are annuals (except for cape gooseberry which is an annual)
2. Fruit plants are asexually propagated	Mostly sexually propagated
3. Training and pruning and are required seasonally	training and pruning not required
4. Plants are generally woody	Plants are generally non-woody
5. Only fruit is edible but sometimes false fruit is also edible	All parts of the plant are edible
6. Mostly consumed raw	Generally consumed after cooking

- Half of the upper 80cm (32inch) of the soil surface has material that is more organic, mineral soils are all other soils.
- The first transgenic plant experiment in the field was started in 1995. In this experiment, *Brassica juncea* plants containing bar gene regulated with plant-specific constitute promoters and linked with Barnase and Barstar genes regulated with floral tissue-specific promoters were planted at Gurgaon (Haryana) India.
- Transgenic rope seeds (*Brassica juncea*) contain Barnase which produces male stencil seeds and Barstar genes (Sterility restorer)
- Indian experiments for Bt. Cotton at industrial (Toxicology Research Centre Lucknow)
- GURT stands for Genetic Use Restriction Technology. V-GURT / T-GURT – V- variety, T-Trait
- Per capita per day requirement of fruit is120g while for average Indian, it is 40-50g National or International annual per capita need of fruit in 34 kg.
- The national average fruit yield is 8-10 MT/ha which is much lower than the world average of 14-16MT/ha.
- Mango is grown in both tropical and subtropical climates. Grape and peach can be grown in both temperate and subtropical regions.
- The arid region of India occupying nearly 12% of the total land area under fruit cultivation.

- Bangladesh is the leading importer of citrus and apples from India. The UAE is the second-largest importer of other fruits and Kuwait ranks the third in the list of importers from India
- Apomictic fruits – fruits which do not require pollination and fertilization for the setting of fruits.
- Parthenocarpic – A fruit that develops without fertilization.
- Single sigmoid – Apple, pineapple, strawberry, ber etc.
- Double sigmoid – Peach, apricot, plum, cherry, grape, fig, currant.
- Triple sigmoid – Kiwifruit
- Long day plant – Passion fruit
- Short day plant – Strawberry
- Neutral day plant – Papaya, Banana
- In fruit crops, self-fruitfulness occurs in peaches, loquat, phalsa, mulberry, bael, Karonda (Moral group) and some varieties of apple and plum.
- Self-fruitful – Apricot, Peach, Grapes, Pomegranate, Loquat, Phalsa, Aonla, Mulberry, Bael, Karonda, Jamun, Plum (some varieties), Sour Cherry (Morella group) and apple (partly fruitful).
- Self-unfruitful – Almond, Ber, Cherry, Citrus, Apple, (Partly unfruitful) Cranberry, Litchi, Strawberry, Mango (some varieties) Triploid apple, date palm, Papaya, Hybrid origin grapes, olive, walnut, chestnut, peanut, plum (some varieties).
- Partly fruitful – Apple, Pear (some varieties)
- Both self and cross fruit fullness – Apple, cashew, citrus, plum.
- Parthenocarpy – Banana, fig (Poona fig) and pear (some varieties).
- In mango, fruit colour is governed by polygene. Red skin colour is dominant over skin colour.
- In papaya, grey seed coat colour dominates over the black seed coat.
- The red pulp colour in guava is monogenically dominant over which flesh colour and bold seed traits are monogenically dominant soft seeds.

Fruit maturity

- Fruits harvested at dessert maturity develop the best quality (Max TSS of 18° Brix in case of mango.
- In case of Alphonso mango, the fruit skin olive green colour stage is the best stage for export.

Stages of maturity of banana

Maturity	Description
70%	Fingers angular, skin dark green.
80%	Fingers slightly less angular skin dark green.
90%	Fingers turning to be round skin dull
100%	Round fingers earliest top fruit turning greenish-yellow.

Maturity stages of date fruits

Name of the stage	Description
Gandor - Chimri	Fruit green developing up to 13 weeks after set.
Doka – Khalal	Changes to red, yellow or pink, full size attained TSS 40% and respiration high.
Dang – Rutab	Fruit softness, starting tip, weight loss for weeks, sucrose hydrolyses to monosaccharides and TSS 55%.

- For a fruit to reach maturity, several factors affect fruit quality parameters like TSS and acid content.
- Similarly, the statement that calcium increases the firmness of fruit, may or may not be the same as calcium affects maturity.
- Generally, higher the temperature, earlier the maturity. Gulabi grapes mature in 100 days while those in the western India take only 82 days to mature.
- Grapes are harvested earlier on light sandy soil than on clay (heavy) soils.
- Close spacing of hill bananas hastens maturity.
- Duration of maturity is shorter in the main crop as compared to that in ratoon crop in banana.
- Removal of the terminal parts of the inflorescence after fruiting under early maturity in banana which could be as much as 14 days in Dwarf Cavendish.
- Most important of these is CEPA in ethylene chemical.
- Some fruits may not attain maturity even on full orange colour development as in Kinnow.
- Rains decrease TSS in mango, increase pest and incidence, leach every sugar in dates in North India are harvested early to avoid damage. normality
- Gales and Cyclones: Orchards in the coastal belt are harvested early to prevent fruit damage by cyclones.

Criteria of maturity for harvesting fruits

Fruit	Physical change	Chemical
Mango	Olive green colour with clear lenticels shoulder develop, size, specific gravity, days from fruit set.	Starch content, flesh colour
Banana	Skin colour, drying of leaves of the plant, brittleness of floral ends, the angularity of the fruit, days from the emergence of inflorescence.	Pulp/peel ration (1:2) starch content
Citrus	Colour break of the skin from green to orange, size.	sugar/acid ratio, TSS
Grapes	Peel colour, easy separation of berries, characteristics aroma.	TSS 18-22 for TS, 12-14 Bangalore Blue, 14-16 Anab-e-Shahi
Pineapple	Greenish-yellow smooth surface around the eyes bract protection dried cup.	TSS 12-14 and 0.5-0.6% acidity
Apple	Colour and size	Firmness is measured by precursor tester and disappearance of fart taste Iodine test for the starch colour of the seeds
Papaya	Yellow patch or streaks	Jelliness of the seed and seed colour

- Apple: Apple is harvested based on the brown colour of the seed. A starch index of 2 to 5 is acceptable for harvesting (Reading of starch index 0-5).
- Fig – Fruits are harvested when soft and wilt the neck hanging down from their weight, milky latex disappears from the bunch as the fruit matures and ripens.
- Litchi- Optimum fruit maturity is reached in 55 days after fruit set with bright pinkish-red colour and feathered tubercles.
- Papaya – Papaya is harvested at the colour break. It takes 145-165 days to attain the eating stage.
- Pomegranate – Harvesting is done at semi-ripe stage when the skin turns yellowish with a waxy surface.

Genetic resources of fruit crops

India is home to many important fruit species.

- *Artocarpus heterophyllus*
- *Citrus indica*
- *Citrus latipes*
- *Feronia limonia*

- *Garcinia indica*
- *Manilkara hexandra*
- *Mongifera indica*
- *Syzygium cumini*
- *Zigybhess mauoutiona*
- *Musa* spp. (AB, AAB group)
- *Musa flaviflora* is localized to Manipur and Meghalaya.

- Maximum genetic variability of *Musa acuminate* and *M. balbisiana* is found in North- East India.
- Mango – Wild forms of *M. indica* and its allied species viz. *M. sylvatica* occur in the forests of North-East region. The fossil leaf impression of *M. pentandra* has been recovered from Assam.
- Banana – Land races mostly belonging to *balbisiana* (BB group) having resistance to drought and cold. Frost resistant species *M. cheesmani* & *M. velutina* farms have been collected.
- Thompson Seedless, Perlatte, New Perlatte, Beauty Seedless, Delight, Himrod and Early Muscat grape cvs. are from the USA.
- Kishmish Charni and Kishmish Bela are grape varieties from Russia.
- Kinnow mandarin and low chilling peach varieties *viz*. Flordasum and Sunred are from the USA.
- Early maturing Satsuma mandarin comes from Japan.
- Naval mandarins, blood oranges, grapefruits, rough lemon and Ponderosa rootstocks come from the USA.
- Bunchy top resistant banana comes from Australia. Peeling type comosus comes from the USA.
- Ring spot virus-resistant papaya car flora comes from the USA.

Introduction of the gene pool

- Downy mildew of grape from Europe.
- Crown gall and hairy root of apple and pear from S. Lamba.
- Bunchy top of banana from Sri Lanka.
- Canker of apple from Australia.
- Red palm weevil of palm from Egypt.
- Woolly aphid of apple and pear from England.

- The fluted scale of citrus, mango and guava from Australia.
- Apple promising varieties *viz*., Lalambri and Sunchany were released from Jammu and Kashmir. Hybrids such as Ambred, Ambrich, Ambstorking and Ambroyal were developed in Himachal Pradesh.
- Cluster bearing varieties of acid lime *viz*., Promalini and Vikram at Prabhami and Kaspentla at Tirupati have been developed.
- A seedless and thornless acid lime; Kadayan has been identified at Periyakulm. Kagzilim, NS-2, H2S4 are resistant to canker.
- Nagpur Santra 182 having negligible seeds was identified in Vidarbha.
- Examples of date palm varieties from North-Western India: Halawy and Bahee for dessert, Khadrarey, Medjool and Shamrem for Chhuhara (Dry dates) Zahidi and Halawy for pind Khajoor (Soft dates)
- Guava Safed Jam and Kohir Safeda are suitable for juice making and are considered drought tolerant.
- Indian gooseberry (Aonla or amla) NA7 is prolific and precocious.
- Mango Lal Sindhri (*Amrapali* × *Sensation*) is resistant to powdery mildew.
- Papaya: - A bisexual cultivar Coorg Honey Dev was released by IIHR Bangalore.
- The mulberry tree is grown in the subtropical, tropical as well as temperate regions of India.
- Central Sericultural Research and Training Institute is located at Sreekandapuram (Mysore).
- Apple: Tropical Beauty (USA), HDP.
- Apricot: New Castle (USA).
- Japanese persimmon: Cultivar flat seedless and Hyakumo have been introduced from Australia.
- Olive: A few selected cultivars *viz*., Cornicobra, Mission, Agopa, Lessino, Cannina and Manola were introduced from Iran.
- Peach: Top Red, Halford, Desert Gold, Sunland, Sun Prince, Florida Prince, Flordaking, Tropic Sweet, Flordastar and White Glory from USA. Chaju from Bulgaria, Sunhavan, Red Haven and May Fire from Australia and Okabo Yamyang, Miropwase and Mardang from Korea.
- Pecan: Mohan from the USA
- Sour cherry: Early Richmond, Nonstar from the USA.
- Sweet cherry: Emperor Francis, Lambert and Napolean from the USA.

Fruit Ripening

- Physical and chemical changes that occur in a ripening fruit mainly relate to the hydrolyzing of polysaccharides to sugar and decrease in acid.
- Besides these changes, oxidative phosphorylation is the source of energy for transcription and translation that are active during ripening.
- **Climacteric ripening:** Climacteric is defined as a period of the ontogeny of fruit during which a series of biochemical changes are initiated by the autocatalytic production of ethylene making the change from growth to senescence involving an increase in respiration leading to the ripening of the fruit.

  Examples: Actinidia, apricot, breadfruit, ber, Chinese gooseberry, feijoa, fig, guava, jackfruit, papaya, passionfruit, peach, pear, persimmon plum, sapodilla etc.
- **Non-climacteric ripening:** These fruits show neither a rise in respiration nor an associated production of ethylene during the ripening process and are hence called non-climacteric.

  Examples: Blackberry, cocoa, cashew apple, cherry, cranberry, grape, grapefruit, java plum, lemon litchi, mountain apple, orange olive pineapple, rose apple, strawberry, surinam, cherry, tamarillo, fig etc.
- Treatments that decrease the level of ethylene in fruits, also lead to their ripening.

  i) Hypobaric conditions are those where in partial pressure of oxygen is maintained at normal atmospheric level but the internal ethylene level is reduced.

  ii) Storage at low O2 concentration which inhibits the ethylene synthesis.

  iii) Storage at high CO2 concentration prohibits ethylene action.

  iv) Treatments with rhizobia in a bacterial product as well as components such as amino-ethoxy vinyl glycine (AVG) and aminooxy acetic acid (AOC) inhibits ethylene synthesis.
- In banana, the critical temperature is reached between 10 and 15°C.
- In Cox Orange pippin apple, the critical temperature is as low as 3°C.
- Raising the level of CO2 and decreasing the O2 level in the storage atmosphere delays ripening.
- ABA and ethylene are important promoters of ripening.
- Ethylene accelerates the degradation of chloroplasts, enhances the biosynthesis of carotenoids and anthocyanins, hence causing softening.

- One of ABAs commonly arises late during the fruit development or ripening.
- Methionine may be a preserver of ethylene.
- Banana fruit bunch matures 90-150 day after the inflorescence emerges from the psoas stem.
- Ripe banana pulp (dry matter) contains 5 to 15% starch and 70 to 80% free sugars.
- Esters import the characteristic banana aroma.
- In mango, acid central varies from 4 to 5% at harvest maturity to 0.1 to 0.5% in ripe roots.
- Protein content in mango varieties varies from 0.5-1.0%.
- Carotenoids and Xanthophyll are the predominant pigments of ripe fruits.
- Carotene constitutes the major carotenoid pigments in unripe (37%) and ripe (50%). Papaya follows a double sigmoid growth.
- Despite containing little starch, papaya fruits show climacteric respiration pattern during ripening.
- Ripe papaya flesh gets its color from the carotenoid pigments of which the most abundant is cryptoxanthin.
- In pineapple, a proteolytic enzyme Bromelain, present in the fruit and stem decreases with the development of fruit.
- AIS – Alcohol insoluble solids.
- Sapodilla seed yields 20% oil.
- Total acidity expressed as malic acid ranges from 0.1 to 0.09 % Vitamin C – 0.1-11.9 mg/100g of ripe.
- Grape growth follows a double sigmoid pattern.
- Grape colour of the berry is due to anthocyanin pigment.
- Grape is a non-climate fruit. It does not exhibit development of colour or taste after harvest.
- Grape tartaric and malic acids account for 90% of the total acidity in berries.
- White grapes contain phonetic compounds in a lesser amount than black grapes.
- Jackfruit requires 100 to120 days to develop into fully mature syncarpous fruits.
- Jackfruit sugars comprise fructose (1.74%), glucose (5.96%) and sucrose (6.9%).

- In apple, flesh firmness ranges from 12 to 22 lbs/sq. in and starch index from 0.5 to 5. These two criteria are used for commercial harvesting.
- Date palm exhibits 5 stages of maturity. **Hababouk** – from pollination to 4 weeks.
  - Chimri or Gandora – green colour up to 13 weeks after fruit set.
  - Khalal or Doka – increase in fruit weight for 4-5 weeks.
  - Rutab or Pind – loss of weight for one week. Jamor or Dong- loss of fruits weight for one week.
- Fruits are edible during the last three (Doka, Pind, Dang) stages.
- In the Doka stage fruit is hard yellow red or pink and weighs 10-15g fruit with over 40% TSS and moisture content between 45-55%.
- Guava: The ascorbic acid content of guava reaches a maximum in green.

Packages	Fruits
Gunny bags	Ber, Lemon, Lime, Mango (raw) sand pear, sweet orange. Bamboo and reed Grape, guava, mango, papaya, basket.
Earthen pots	Custard apple, grape.
Wooden boxes	Apple, apricot, cherry, litchi, mango, mandarin, pear, plum, sapota. CFB cartons Apple, cherry, grape, pomegranate and fruits for export.
Rigid Plastic Crates	Loos fruits for the public distribution system. (RPC)

- Watermelon rind is at present used for the preparation of 'Tutti Fruity'.
- The seed of wild pomegranate is sundried to make *Anardana* which is used for garnishing currish.
- Possible by-products from wastes of the fruits processing unit.

Fruits	Waste%	Nature of waster	By-products
Mango	27-50	Peels, stones and waste from pulping machines	Starch, fat, pectin, vinegar, syrup, alcoholic beverages
Pineapple	50-60	Peels, cores & trimmings	Vinegar syrup citric acid, candied cores, stock feed
Grape	5-10	Stem, seeds, seed hulls	Cream of tartar seed oil, Tannen from culls vinegar, wines pectin, stock feed
Apples	60-70	Pomace, cores, cull fruits	Pectin, cider, beer, vinegar soft drinks, jelly base
Citrus	50	Peels, pomace and seeds	Pectin, essential oil, seed oil etc.
Banana	24-46	Peels	-
Peaches, apricots		Fruits pitches	Fixed oil, bitter
Almonds cherries		Kernel	Almond oil, stock seed

- Quince (Bihi) seeds have been used in the preparation of indigenous medicines.
- Pectin is isolated from Gellan Gum (gg) and the characterized latex is utilized in the chewing gum industry for preparing chicklet.
- Second drop / June drop: Wind, low atmospheric humidity, low soil moisture and high temperature.
- Pre-harvest drop / September onwards: deficiency of auxin causes this drop.

Control

- Sweet orange – June drop
- PGR – 2,4-D, 2,4,5 T – 10-20ppm and GA- 20,50,75ppm
- Pre-harvest:
    - 2,4,5-T-30ppm followed by 2,4-D-5ppm
    - Alcar & ccc – 250ppm
    - 2,4-D-15ppm + Bordeaux mixture + 0.5% Zn
- Mandarin: 2,4-D, 5-20ppm and 2,4,5-T-20ppm
- Kinnow – NAA – 15ppm, 20ppm – 2,4-D; Lime & Lemon – NAA 5ppm, 10ppm GA.
- Nutrient element: 7% ZnSO4 – Kinnow 5% ZnSO4 + 15ppm 2,4-D + Borders mixture – sweet orange.
- In citrus, all the known forms are diploid except *C. latifolia* which is triploid.
- The related genera of the subfamily Auranlodcae indigenous to India are: *Aegle mandos*, *Feronia limonia*.
- Hill lemon, Galgal and Kil Kil – *C. limon* var. decumam.
- Small fruited mandarin – *C. reshmi*
- *C. reticulate* occupies the first position among Mandarins or lose skinned oranges
- Sweet orange is the second most important fruit grown in India.
- Sathgudi of Andhra Pradesh has large, seedy fruits with abundant juice and is believed to have been introduced from Bateria (Indonesia).
- Cultivars *viz.*, Jaffa, Valenta and Honlin have been introduced from Florida.
- Lime and lemons are cultivars highly tolerant to leaf miner and citrus canker but the fruit has less acceptability as table fruit and is mainly used for processing.

- Rough lemon (Rootstock), about 75% of mandarins and sweet oranges are propagated on various strains.
- Leu foliate orange: It is an exotic collection having non-edible fruit but has proved to be a donor for phytophthora and nematode resistance in hybrids despite having performed well as a rootstock for mandarin in acidic soils.
- Sweet lime exhibits susceptibility to Tristeza and phytophthora.
- Citrange/citrumelo: this exotic collection of men made hybrids are promising as rootstock for mandarins.
- *C. macroptera* most (Sat kara) is frost resistant and may be used in the breeding programme as well as a rootstock.
- Citrus seeds are recalcitrant and lose viability within 2.4 weeks of Maxi polyembryony acid lime (98%).
- The productivity of citrus trees in India is very low with an average of 7.62 tones/ha as compared to 17-30 tones/ha in Spain, Italy and Japan.
- Sweet oranges var are tolerant of Tristeza and exocomets but are highly susceptible to Greenburg and moderately susceptible to phytophthora.
- Mandarin and tangerine var are highly tolerant of Tristeza but are susceptible to Greenburg.
- The var know is a hybrid of (*C. nobles* × *C. delicious*) is susceptible to canker but free from granulation.
- Grapefruit variety is susceptible to Tristeza and greening pummelos are tolerant of greening.
- Two Kagzi time varieties (Pramalini and Vikram) have been developed by donal selection from (MAU Parbhani).
- Thornless and seedless selections (Chakradhar) have been made from Kagzilime. 60-60% juice.
- Greening, sweet orange or tangelo seeding are used as indicator plants.
- Sour orange has been the most dominant rootstock in the viticultural world. It possessesesa uniform quality and is regular and early bearing.
- citrus micropropagation is mostly achieved using meristem and axillary buds.
- Anther culture has been successfully carried out in *C. aurantium* and *C. curatifola*.
- Triploids of *C. grandis* have been produced from endosperm culture.
- Monoembryonic varieties (Clementine cultivar *C. reticulata*, *C. limon* and *C. grandis* do not produce nuclear embryos.
- Vidarbha region of Maharashtra is the largest citrus cultivation area in India.

Indicator plants

- Acid lime – Tristeza
- Sweet orange – greening bacterium
- Etrong citron – exocortis
- Sweet orange and Chenopodium quinoa – ring spot.
- Budding is performed in the last week of the first week of February.
- Dwarfing rootstock – Trifoliate oranges, Citrange, Dragonfly
- Vigorous rootstock – Rough lemon Sour orange
- Only 1-2% flowers become harvestable fruit.
- Although lime and lemons flower throughout the year, another citrus usually flowers twice a year.
- The flowers of 'Washington Navel' and 'Satsuma' have no viable pollen. They are male sterile, only seedless fruit.
- In sweet orange, *Alternaria citri* and *Colletotrichum gleospolliodes* are largely associated with fruit drop. The same can be controlled by copper fungicide.
- It causes up to 22% preharvest drop in the orchards having 10% dead twigs of the canopy.

Seed

- Food storage organ – Dicot – Cotyledons Food storage organ – Monocot – endosperm.

  (Except caster, brassica in which food stored in endosperm)
- In gymnosperms, storage is the haploid female gametophyte. Double fertilization (triple fusion) does not occur in gymnosperms.
- Selected enzymes synthesized during germination.

Enzyme	Function
Amylase	Starch hydrolysis to sugar.
Proteases	Protein hydrologist to amino acid.
Lipase	Lipid (oil) hydrolysis to fatty acid & then into sugars.
Gluconases	Cell well-degrading enzymes.

- The seed germination does occur at moisture below 40-60%.
- The optimum temperature range for the seeds of most of the fruit plants is between 25-30°C.

- Germination of seeds is augmented by Red light (660-730nu), whereas the far red (730-800 nu) light inhibits germination.
- Vegetative apomixes, in some cases, vegetative buds or bulbils are produced in the inflorescence in place of flowers in inflorescence, which subsequently develops into a new plant.
- The phenomenon in which two or more embryos are produced in a seed is called Polyembryony *e.g.*, Citrus, Jamun, Mango.
- Stratification requirement of fruit crops:

Fruit crop	Optimum temp (°C)	Period of treat (days)
Apple	3	70-80
Pear	3	70-80
Plum	3	70-80
Peach	5	60-100
Grape	5	80
Cherry	3	120-150 days

- In mango, Anupam has been recommended as an inter-stock for Amrapali, grafted on Mallika.
- Root apexes are an important site of ABA synthesis. ABA levels are higher in dwarfing rootstock.
- Auxin – Bud and leaf, shoot apexes
- GA31 Cytokinin, ABA – Root, root apexes
- Ethylene – Fruit
- The tissue culture material is released within 6 hours at the point of entry.
- Greenhouse cultivation in Holland is very old. Almost all greenhouses in Holland use glass as the glazing material.
- Italy has the second largest area under fruit cover in Europe, China & Japan are the largest users of protected cultivation in the world.

Autonomous body

1. National Horticulture Board, Gurgaon, NHB-1984
2. Coconut Development Board, Cochin – CDB
3. National Communication on use of Plastic in Agriculture, NCPA, New Delhi.
    - High light intensity induces dwarfing in the plant.
    - High light intensity leads to greater flowering than low intensity.

- The transformation of vegetative buds into flower buds is called flower bud initiation. In other words, the conversion of vegetative phase into reproductive phase is known as fruit bud differentiation.
- A particular C: N ration (3:2) in shoots is essential to produce flowers.
- Fruit bud differentiation in banana takes place five months after planting of the suckers.
- Temperatures lower than 20-25°C favour greater accumulation of carbohydrates and thus forcing the flower bud formation.
- Flowering in the long day plants and short-day plants is accelerate by excess CHO and nitrogen respectively.

Bearing habit

- Apple and pear: Mixed bud which bears terminally occasionally on the shoot and principally on spurs.
- Ber: Mixed flower bud that bears on the current season growth laterally on the shoot.
- Cashew nut: Flower bud is mixed and bears on one-year-old shoot / last season growth.
- Cherry: Flower bud simple (FBS) and bears on spurs laterally.
- Citrus: FBS and bears laterally.
- Coconut: FBS and bears laterally.
- Date palm: FBS and bears laterally.
- Grape: FBS bears on spurs laterally as well as on current season growth.
- Guava: Bears on terminal spurs, flower bud mixed with flowering shoot bears lateral inflorescence.
- Pomegranate: FBS bears on lateral axillary inflorescence.
- Mango / Loquat: FBS bearing terminal on CSG.
- Papaya: Current season growth laterally on the shoot.
- Peach: Flower buds simple, bear laterally on last season growth or one-year-old shoot.
- Sapota bears fruits all along the length of old wood in the leaf axil.
- Walnut bears fruits on the current season growth either terminally or laterally.
- Jackfruit is a stem or branch bearing (Cauliflorous) fruit.

- In developing fruits, the presence of tryptophan has been reported in large quantities and the seed possibility supplies the factor responsible to convert it into auxin.
- Auxin application can control a drop of fruit arising from self-pollinated flowers known to be more susceptible to drop.
- Seed lessness is observed in grapes, fig, mandarin, orange, sweet orange, and guava.
- Citrus is the second most important fruit crop after grapes in terms of total area under cultivation and production in the world.
- In India, blight in Florida, decline in Brazil and declinate in Argentina.
- In India, citrus decline was first reported in MP as early as in the eighteenth century.
- The best pH for citrus soils ranges from 5 to 6.
- Excess calcium carbonate is an important contributing factor for decreased availability of iron and zinc.
- Iron plays an important role in the electron transport system of the plant.
- Excess of N, directly or indirectly influences the availability of Cu, Zn, Mn, Mo, P and other elements.
- Continuous use of NPK in the inorganic form leads to a deficiency of Mg, Cu and Zn.
- Excess of Iron and salinity and high uptake of P and Mn cause dieback.
- Excess N promotes vegetative growth and delays fruit maturity.
- The excess Zn also induces leaf burn defoliation and twig dieback.
- Excess Zn shows iron chlorosis.
- The most predominant nematode in the growing belts is the citrus nematode *Tylenchulus semipenetrans*
- Use of *Trichoderma* culture also controls *Phytophthora*.
- Rough lemon is tolerant to Citrus Tristeza Virus but is highly susceptible to *Phytophthora*.
- Granulation was first reported from California and in India from Delhi and Abohar.

Deficiency symptoms	Deficient nutrient
Leaves are light green to light yellow	Nitrogen
Leaves are dark green, red and purple the lower leaves may be yellow, Turin greenish, brown and dusty black	Phosphorus
The lower leaves mottled or chaotic	Magnesium
Dead spots on lowers leaves, tips	Potassium
Dead spots on lower leaves in large area	Zinc
Terminal buds of the found leaves, pooked, Which dies soon at tips and margins	Calcium
Terminal buck and young leaves light green	Boron
Young leaves are chlorosis	Copper
Spots of the dead on young leaves	Manganese
Chlorosis of young leaves with the vein, light green	Sulphur
Chlorosis is young leaves, veins green	Iron

2

Post-Harvest Management and Marketing

- Agriculture – 26% GDP, Processed food industry – 6.3% GDP.
  - i) Simple fruits: Fruits which develop from a single ovary of a single flower e.g., Lemons, citrus, peach, apple, pear etc.
  - ii) Aggregate fruits: Fruits which develop from many ovaries of a single flower e.g., Raspberry, strawberry, blackberries.
  - iii) Multiple fruits: Fruits which develop from many ovaries of many flowers e.g., Pineapple.
- Pre harvests spray of calcium and brown on fruits and results in improving firmness.
- Worldwide postharvest fruits and vegetable losses are as high as 25-40.

Postharvest horses	% of total productions
Papaya	40-100%
Citrus	20-95%
Banana	20-80%
Avocado	43%1
Stone fruits	28%
Grape	27%
Apple	14%
Overall	30-40%

Reasons for high postharvest losses

a) Unfavorable climate:

b) Poor cultural practices:

c) Improper storage conditions:

d) Inadequate handling during transportation:

e) Inadequate: facilities available for processing – India-2% process / developed countries – 50-70%.

f) Government policies:

g) Education and resources:

h) Transportation facilities:
  - 60 Agri-export zones.
  - Level of processing in fruits and vegetables in different countries.
    - UK and USA – 80%
    - Malaysia – 80%
    - Thailand – 30%
    - India – 1.8-2.2%
  - Fruit Product Order (FPO) – 1955.
  - India ranks 6th in poultry production in the world.

Maturity and Ripening

- Fruits and vegetable can be divided into 3 stages.
- Growth: This generally refers to cell division, enlargement and differentiation ultimately giving a particular size, weight and volume to the commodity.
- Maturation: It is initiated before the cessation of growth and lasts until the onset of senescence.
- Senescence: The stage when anabolic (synthetic) processes almost terminate and catabolic (derivative) processes are initiated and speeded up causing ageing and finally death of tissues.
- Ripening is the stage that begins with the last stages of maturation and lasts till the beginning of senescence. Ripening marks the completion of the development phase.

Important physic-chemical changes occurring daring, ripening.

- Maturation of seeds
- Changes in fruit skin colour/development of carotene.
- Loss of chlorophyll
- Development of anthocyanins
- Foundation of abscission layer.
- Changes in respiration rate
- Changes in ethylene production
- Softening
- Changes in carbohydrates
- Flower changes
- Organic acid

Respiration for synthetic reaction

$C_6H_{12}O_6+6C_2 = 6CO_2 + CH_2O$ + energy

2 types

i) Aerobic respiration (occurs in the presence of O_2)

ii) Anaerobic respiration (occurs in the absence of O_2)

Respiration rate: Rate of consumption of oxygen or the evolution of CO_2 from a commodity is called a respiration rate. (ml. CO Kg-1h-1)

Classification of fruits, based on respiration

Category	Respiration rate	Crops (at 5°C) (mg $CO_2Kg^{-1}h^{-1}$)	Crops at 15°C (mg $CO_2Kg^{-1}h^{-1}$)
Very low	Less than 5	Dried fruits and nuts, dates	Maybe nuts
Low	5-10	Apple, citrus, grape, kiwifruits	
Moderate	10-20	Apricot, banana, cherry, peach, plum, fig	Grapes, lemon, orange
High	20-40	Avocado, blackberry, raspberry, strawberry	Apple
V. High	40-60	-	Banana (green) peach
Extremely high	More then 60	Mushroom	Banana (ripe) strawberry

Growth Curve

- a-b – Cell division, very small size of fruits on the tree.
- b-c – Cell enlargement, increase in size parameter *i.e.*, length, breadth, diameter, weight, volume etc. of the.
- c-d – Negligible weight gain in fruit, the fruit matures and entry into ripening and senescence at the end.

Climacteric Respiration Curve

i-ii) Sharp decline in respiration.

ii-iii) Respiration rate becomes steady in maturing fruits.

iii-iv) Respiration rate gain rises (climacteric rise) and reaches a maximum (climacteric maximum or peak). It is very difficult to control the respiration rate in the stage of fruit growth. Storability reduces very fast due to fast ripening and early entry into senescence.

iv-v) Respiration rate again drops down (post-climacteric) when fruits enter into senescence.

Non-climacteric Respiration Curve

- A-B – Sharp decline in respiration.
- B-C – Respiration rate becomes steady in fruits. There is no climacteric rise. Fruits are required to mature on trees.
- Respiration Quotient (RQ): It may be defined as the ratio of carbon dioxide produced to that of oxygen consumed during respiration.
- RQ = CO2 produced (in ml) / O2 consumed (in ml) Complete oxidation of glucose RQ – 1.0
- Complete oxidation of malate RQ – 1.3 Complete oxidation of stearic acid RQ –0.7
- Low RQ would indicate the metabolism of fats and high RQ would indicate the metabolism of organic acid *i.e.*, more organ per carbon atom.
- Climacteric fruits: A sharp increase in respiration is shown by the increase in the production of CO2 and decrease in internal O2 concentration.

E.g., Apple, Apricot, Avocado, Banana, Blueberry, Breadfruit cherimoya, Feijoa, Fig, Guava, Jackfruit, Kiwifruit, Mango, Muskmelon, Peach, Papaya, Passion fruit, Pear, Persimmon Plum, Sapota, Tomato, Watermelon.

Non-climacteric fruits produce much lesser amounts of ethylene than climacteric fruits.

E.g., Ber, Blackberry, cashew apple, cherry, cucumber, grapes, citrus, litchi, loquat, olive, pineapple, pomegranate, strawberry.

Climacteric	Non-climacteric
1. Fruits exhibit respiration climacteric (Pronounced increase in respir ngrespiration coincident with ripening farming a pearl.)	1. Fruits do not exhibit a respiratory the climacteric.
2. Coincident with ripening, the fruits produce larger amount of ethylene.	2. No such relationship exists in a much these fruits.
3. Fruits may ripen on and of the tree. *e.g.*, Apple, guava, papaya, mango	3. Fruits ripen on the tree only. *e.g.*, Citrus, grapes, litchi

Factors affecting respiration rate

1. Temperature

   Van't Hoff, a Dutch chemist reported that the rate of a chemical reaction almost doubles for every 10°C rise in temperature.

   Van't Hoff temperature quotient, $Q_{10}=(r_2/r_1)10(t_2\text{-}t_1)$

   Where: r_2 and r_1 – are the respective rate of reactions at temp. t_2 and t_1 in °C.

Q_{10} values are the highest at 1-10°C temp. and can be as high as 7 but at higher temp, Q_{10} values are between 2-3.

2. Atmospheric composition
3. Effect of ethylene:

   Effect of applied ethylene on Climacteric and Non-climacteric fruit.

Climacteric fruits	Non-climacteric fruits
Ethylene applied @0.1-1µl./Lit. one day is not hasten full ripening.	Ethylene applied @0.1-1µl/Lit for one day sufficient to hasten full ripening. merely increases the respiration and does
The magnitude of climacteric rise is 0.1 as well as 1 or 10µl/Lit.	The magnitude of climacteric rise is 10 or 1000µl/Lit. and.
The rise in respiration is a response to applied anyone more than once.	Rise in respiration is a response ethylene to applied ethylene occurs more than once

Ethylene Biosynthesis

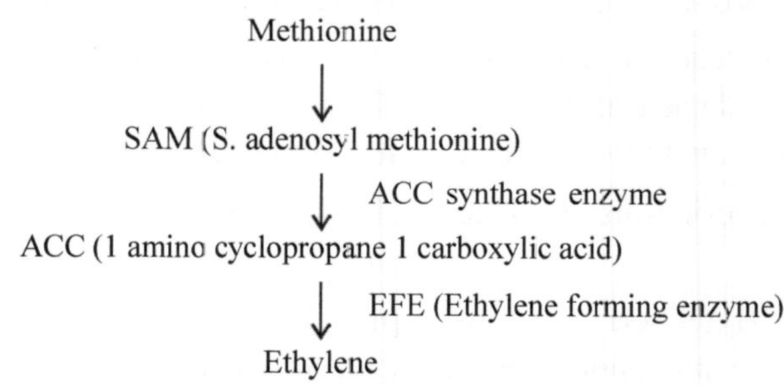

Classification of fruits based on the ethylene Reduction rate

Category	Ethylene evolution	Crops at (20°C) rate (µl ethylene kg-1h-1)
Very low	Less than 0.1	Citrus, grapes, ber, strawberry, pomegranate.
Low	0.1-1	Blueberry, cranberry, olive, persimmon, pineapple, lemon, lime orange, raspberry, watermelon.
Moderate	1-10	Banana, fig, guava, mango.
High	10-100	Apple, apricot, peach, plum, pear, cantaloupe, feijoa, kiwi fruit, papaya.
Very high	more than 1000	Apple, nectarines, cherimoya, passion fruit, sapota.

Biochemistry of Respiration

Different carbohydrates get converted into glucose or other intermediate products of glycolysis.

- Vital heat: Vital heat is the energy released in the form of heat during respiration.
- Methionine is the precursor in ethylene biosynthesis.

Quality management for fresh marketing

- Harvesting: Harvesting is detaching a commodity from the point of its origin. This point of origin may be an above-ground plant part *i.e.*, shoot *e.g.*, Apple, etc. and underground plant part *e.g.*, potato etc.
- Maturity: It is a stage of fruit growth preceding ripening. Physiologically, the meaning of 'mature' and 'ripe' are not the same although most common people use these two terms as synonyms.
- Physiological maturity is the stage at which a plant or a plant part continues ontogeny even if detached.
- Horticultural maturity in the stage at which a plant or a plant part possesses all the pre-requisites for utilization by the ultimate consumer for a particular purpose.
- Maturity indices are the factors based on which it is decided whether the horticultural commodity is ready for harvesting or not.

Physical methods: (Maturity)

Method	Crops
Fruit retention strength	Apple Fruit size & surface morphology
Netting	Melons
Cuticle formation	Grapes, tomato
Development of wax	Plum, apple
Specific gravity	Cherries
Flesh firmness	Apple, pear temperate stone fruits.
(T stage)	Apple
Tenderness	Pea, beans
Juice content	Citrus
Oil content	Avocado
Development of abscission layer	Apple, Feijoas Fullness of cheeks Mango

Chemical Methods

Methods	Crops
TSS	Apple, plum, pear, apricot
Titratable acidity & TSS: acid ratio acid ration	Pomegranate, Citrus, Papaya, Grapes Sugars & Sugar: Apple, pear, stone fruits, grape Starch Apple, pear
Astringency/tannins	Persimmon, dates, walnuts, pecan nuts.

Quality management for fresh marketing

- Freezing injury: (Less than – 0°C)

- Chilling injury: This type of injury occurs in subtropical or tropical fruit crop.
- Heat injury: Exposed to direct sunlight or excessively high temperature.
- Irradiation can also be used to control diseases, disorders and pastes.
- Irradiation can also be used to control disease, disorder and pasts.

Crop	Control of Disease/disorder	Min. dose (KGy)
Apple	Scald / brown core /	1.5
Apricot, peach	Brown rot	2
Banana	Ripening inhibition	0.30-0.35
Lemon	Penicillium rot	1.5-2.0
Mushroom	Inhibitions of stem growth and cap opening	2
Orange	Penicillium rate	2
Papaya	Disinfestation of fruit fly	0.25
Strawberry/grape	Grey mould	2

Relative tolerance of fresh fruits to Irradiation below 1 KGy

Low	Avocado, grape, lemon, lime, olive.
Medium	Apricot, banana, cherimoya, fig, grapefruit, loquat, litchi, orange, passion fruit, pear, pineapple, plum.
High	Apple, cherry, date, guava, muskmelon, nectarine, papaya, peach, rambutan, raspberry, strawberry.

- 500 ppm of benomyl in the water at 50-55°C for 2-15 min is effective for controlling anthracnose in mango without damaging the fruit.
- Fumigation – Sulphur dioxide – a postharvest disease of grape. Waxing/ coating: Paraffin wax, Carnauba wax, bear wax.

Cooling methods

- Room cooling: Banana, coconut, lemon, orange, pineapple.
- Forced air cooling: Banana, Barbados Cherry, Berries, fig, grape, guava, litchi, mango kiwifruit, mushroom oranges, papaya, passion fruit, pineapple, persimmon, pomegranate, quince, sapota, strawberry.
- Hydro cooling: Kiwi fruit, orange, pomegranate.
- Ice cooling (Package icing):
- Vacuum cooling: Mushroom
- Transit icing: All fruits and vegetable.

- Washing, Cleaning and Trimming: Washing was done in plain water or chlorinated water or fungicides *i.e.*, (Diphenylamine) (0.1-0.25%) and ethoxyquin (0.2- 0.5%) as a postharvest dip.
- Size grades of apples

Grade	Min. Dia (mm)
Super large	85
Extra large	80
Large	75
Medium	70
Small	65
Extra-small	60
Pitoo	55

- Size grades of apricot

Special	:	42
Grade I	:	36-42
Grade II	:	<36

- Size grades for Malta oranges

Extra special	3.25 (in)	8.25 (cm)
Special	3.00 (in)	7.62 (cm)
Good	2.75 (in)	6.98 (cm)
A	2.50 (in)	6.35 (cm)

- Size grades for Malta oranges

Grade	Size cm	No of fruits / 10 kg
Small	5.0-5.5	115
	5.0-6.0	98
Medium	6.0-6.5	84
	6.5-7.0	76
Large	7.0-7.5 >7.5	64

- Size grades for Pomegranate

Grades	Fruit Weight (g)	Fruit Characteristics
Super	>150	Attractive, very large, dark red coloured without blemishes.
King	500-750	Attractive, large, without blemishes
Queen	400-500	Large medium, attractive without blemishes
Prince	300-	400 medium attractive, without blemishes
12-A	250-300	Small fruits, may have 1-2 spots
12-B	<250	Very small size

- Size grades for pineapple

Grade	Fruit Weight (g)
A	>1500
B	1100-1500
C	800-1100
D	<800
Baby	550 (Approx)

- Size grade for mango

Grade	Fruit weight (g)
Category I	200-250
Category II	251-300
Category III	301-350

- Size grades for litchi

Grade	dia (mm)
Extra class	33
Class-I	28
Class-II	23

- Size grades for guava

Grade	Weight (g)	Dia (mm)
A	>350	>95
B	251-350	86-95
C	201-250	76-85
D	151-200	66-75
E	100-150	54-65
F	61-100	43-53

- Size grades for papaya

Grades	Weight (g)
A	200-300
B	301-400
C	401-500
D	500-600
E	601-700
F	701-800
G	801-1100
H	1101-1500
I	1501-2000
J	>2001

Difference between sorting and grading

Sorting	Grading
1. Undesirable types of fruits such as diseased, deformed are removed.	1. The fruits are categorized according to damaged the difference in their weight, size colour, maturity etc.
2. Done to reduce the spread of infection to other fruits.	2. Done to fetch a better price in the market.

- Curing – Heat (~30°C)
    - Good ventilation
    - Low humidity

Packaging

- Pomegranate: Corrugated fibre box (CFB)
    - Brow coloured 3 ply CFB boxes are used for local markets. White coloured 5 ply CFB boxes are used for distant markets.
- Packing of Pomegranate:

Grade(s)	Size of CFB box (cm)
Supper and king	32.5 × 22.5 × 10
Queen	37.5 × 27.5 × 10
Prince and 12A	35 × 25 × 10

- Citrus: CFB boxes of 3 ply – 5 ply
- Box size - 49.5 × 29.5 × 17.5 cm having 5% ventilation with 10 kg capacity
- Grapes: Table grapes for export are packed in 5 ply corrugated boxes of size 500 × 300 mm and 5 kg capacity.
- Pineapple: Bamboo baskets (20-25 kg) lined with a paddy-5 to all.
- Banana: Bunches are paddy with banana leaves.
- Stone fruits: small packs of 4-6 kg capacity in CFB.
- Storage of fruits and vegetable:
- Storage: It can be defined as keeping the commodity fresh or processed in safe condition with minimum deteriorative change for later use.

General recommended storage condition for fruits

Crops	Temp-°C	RH	Duration
Apple	0-4	80-90	4-8 months
Apricot	0-2	80-85	1-2 weeks
Banana	12-13	80-90	1-2 weeks
Cherry	0-2	85-90	2 weeks
Dates	7-8	85-90	2 weeks
Fig	0-2	85-90	4 weeks
Grapes	-1.1-2	80-95	3-8 weeks
Guava	-7.2-10	85-90	2-3 weeks
Jack fruit	11-13	85-90	1-5 months
Kiwi fruit	-0.5-0	90-95	3-5 months
Lemon	8-10	85-90	1-6 months
Lime	8-10	85-90	3-6 weeks
Litchi	0-2.1	85-95	3-10 weeks
Mandarin	5-8	85-90	10-14 weeks
Mango	8-12.8	85-90	2-7 weeks
Papaya	7.2-10	80-90	1-2 weeks
Passion fruit	7.2-10	80-90	4-5 weeks
Pear	0-1	85-90	3-6 months
Peach	0-3	85-90	2-4 weeks
Persimmeri	0-2	85-90	3-6 weeks
Plum	0-2	85-90	4-8 weeks
Pomegranate	0-2	80-90	2-6 weeks
Sapota	3-4	85-90	6-8 weeks
Strawberry	0-2	80-95	5-7 days.

- Zero energy cool chamber (ZECC) is a low-cost storage structure suitable for start duration of fruits.

  Principle (ZECC): (Evaporation causes cooling)

  (ZECC) temp-10-18°C and storage life increases by 2-3 times RH-90- 95%

  Almost all fruits: Apple, mango, plum, apricot, pear, guava, citrus.
- Water the chamber regularly at least 2-3 times a day.
- Always store the fruits in plastic crates/trays only.

  ZECC is not good in July - August because of low evaporation (short duration)
- Walk-in evaporation cool chamber:

  Refrigeration is the process of removing heat from an enclosed, space or a substance.

  1 ton refrigeration = 200 Btu/min = 3.517kj/s = 3.517 KW.

- In controlled atmosphere (CA) storage we reduce the concentration of O_2 and increase the concentration of CO_2 in storage atmosphere to fixed levels to slowdown respiration rate and repairing temperature-0-15°C (O_2-0-10%, CO_2-0-20%)

  O_2-20.9% Atmosphere N2---78.1%

  CO_2-0.03%

Recommended CA condition for fruits

Fruits	Temperature	% O2	% CO2
Apple, apricot, peach, plum, prunus, kiwifruit	0-5°C	1-3	0-5
Strawberry	0-5°C	10	15-20
Persimon	0-5°C	3-5	5-8
Nuts and dried fruits	0-25°C	0-1	0-100
Citrus, mango, papaya, pineapple	-100-5°C	5	5-10
Banana	12-15°C	2-5	2-5
Avocado	5-13°C	2-5	3-10
Orange	5-10°C	10	5
Olive	8-12°C	2-5	5-10

- Modified atmosphere (MA) storage: Removal of oxygen *e.g.*, – Ferrous oxide.
- Hypobaric storage / low pressure storage: Low pressure – 0.2-0.5 atm

  Temperature – 59-75°F or 15-23.9°C O2 reduces from – 21 to 2.1%
- CO2 scrubbers are hydrated lime, aluminum calcium silicate, activated carbon etc.

Packaging of fresh and processed food products

Pre-packaging: Pre-packaging is to warp or pack a commodity in small convenient units before marketing or to package as food or a manufactured well before offering for sale to the consumer.

- Modified Atmosphere Packaging (MAP) Increase shelf-life – 50-400%

  Hygienic pack having good stacking strength.

  CFB stands for corrugated fibre board and the cartons come in 3 ply or 5 ply.
- General principles and methods of food preservations:

Methods of food preservation

1. Asepsis
2. Removal of microorganisms:
   - By washing
   - By trimming
   - By centrifugation – fruit juices, drinking water
   - By sedimentation - fruit juices, drinking water
   - By filtration fruits juices, drinking water beverages soft drink wine beer etc.
3. High temperature:
   - Thermal death time:
   - Thermal death points: microorganisms are killed in 10 minutes.
   - Decimal reduction time: 90% reduction of micro-organic.
   - Pasteurization: temp below – 100°C
   - Sterilization: temp above – 100°C
   - Simmering: Gentle boiling at about 100°C
   - Blanching: boiling water at about - 100°C – 3-5 min.
   - Frying: temp – above 100°C, but the internal temp does not reach 100°C.
   - Roasting: Internal food temp reaches about 60-85°C in meat.
   - Cooking:
4. Low temperature:
   - Freezing: low temp.
   - Sharp (slow) freezing: (-15 to -29°C) temp. – 3-72hrs.
   - Quick freezing: -17.8-45.6°C temp – 3 min.
   - Dehydrofreezing:
5. Drying and Dehydration: Keeping the food in open sun temperatures. – 50-60°C in air, freeze-drying – (lyophilization), keeping the high concentration of sugar – 24-48hrs – hot air (osmatic dehydration)
6. Use of additives (preservatives)

   I- Sugar, soft, oil.

   II- Potassium metabisulphite (KMS) sodium benzoate, sorbic acid etc.
7. Anaerobic conditions: Additions of CO2, Pepsi, Coca-cola, Fanta, Dew.
8. Irradiation: is also called (old sterilization) cobalt-60, cesium-137.

9. Modern methods:
    - High-pressure processing
    - Use of pulsed electric and magnetic fields
    - Use of pulsed light temp.
    - Ohmic heating
    - Ultrasounds
    - Linear induction electron accelerator.

Stout term / temporary preservation methods	Long term / permanent preservation method
Asepsis, anaerobic conditions low temp, pasteurization filtration, centrifugation blanching, sedimentation trimming, washing etc.	Sterilization, canning, drying fermentation, high pressure, processing, aseptic processing packaging, irradiation, use of preservatives etc.

Difference between Pasteurization, Sterilization, Blanching

Pasteurization	Sterilization	Blanching
1. Aims at killing of most but not all of the microorganisms.	1. Aims at killing of all the microorganisms.	1. Aims at inactivation of enzymes rather than killing of microbes.
2. Temp. <100°C	2. Temp. >100°C	2. Temp. 100°C (boiling water)
3. Generally used in fruits.	3. Generally used in vegetable.	3. Used for both Fruits and Vegetable but only solid pieces). we cannot blanch liquids.

Classification of food products based on shelf life

- Non-perishable: Low moisture - <8-10% *e.g.*, sugar, flour, dry beans. (Some months)
- Semi-perishable: Few days to about 9 months *e.g.*, potato, nuts, almonds.
- Perishable: More than a day or two high concentrations of moisture - >70- 80% *e.g.*, milk, meat, fruits, leafy vegetables.
- pH concept: pH is the –ve log anthem of hydrogen ion concentration in a product or logarithm of the number of litters of solution which contains 1g of hydrogen ions.

Classification of food products based on pH (Cameron 1940)

- Low acid foods: pH >5.0 *e.g.*, Peas, beans, corn, meat, fish, poultry milk.
- Medium acid food: pH >5.0-4.5 *e.g.*, meat, vegetables, soups.
- Acid foods: pH – 4.5-3.7 *e.g.*, tomato, pears, figs, pineapple.

- High acid foods: pH <3.7 *e.g.*, Pickle, citrus, juice anola, berries.

  pH 4.5 is regarded as the dividing sauerkraut lime between acid and non-acid foods and pH 7.0 between acidic and alkaline.

- Heat processing Fruits: boiling water – 100°C

  Heat processing vegetable: boiling temp – 115-121.1°C

Tomato products

Products	Min TSS (°Brix)	Remarks
Juice	5	Not more then (0.5%) w/w.
Soup	7	Salt, sugar, dextrose, malic acid, ascorbic acid citric acid and permitted colours (for juice and soup)
Medium	9	
Heavy	12	
Pasta	25	
Ketchup	25	1.0 acidity
Sauces mixed	15	1.2 acidity

Difference between ketchup and sauce

Ketchup	Sauce
1. Prepared from tomato only	1. Prepared from tomato as well as other fruitssuch as pumpkin, chili etc.
2. TSS - 25% (min)	2. TSS - 15% (min)
3. Min acidity – 1.0%	3. Min acidity – 1.2%
4. Thicker in consistency	4. Thinner inconsistency
5. Costly	5. Cheap as compared to ketchup
6. Only red in colour	6. May have red, green or then colour.

Pickles, fruit chutney and sauces

Some pickles like 'Sauerkraut' are not very common in India but are very popular in European countries.

- The lactic acid bacteria are salt-tolerant and they can flourish in a brine of about 8-10% salt content.
- Lactic acid fermentation:

  Carbohydrates → Homofermentative → Lactic acid

  Carbohydrates → Heterofermentative → Lactic acid, acetic acid, CO_2, Ethanol.

  Gherkins: Small-sized cucumbers are brined for pickling.
- Sauerkraut – Fermented cabbage – (<2 - >3% salt)
- Salt in pickles 12%

FPO specifications

1. Pickles in vinegar: Not less than – 2% acidity in fluid portion as citric acid.
2. Pickles in citrus: <12% salt Juice and brine: <12% acid

- Chutney: TSS not less than – 50% Fruit part not less than 40%
- Sauce: TSS-15%, 1.2 acidity
- **Black neck** is an important problem in sauce and ketchup.

Sugar >66% (Jam, Jelly, Marmalade and Preserve)

Jam: 45% parts of pulp. pH – 3.35-3.70. 55% parts of sugar

Pectin – 0.5-1.0%

68% TSS Acidity 0.5-.07%

Jelly: Transparent - 33.38% water

- 60-65% sugar, fruit acid – 1.0% (optimum 0.75%)

- 65% TSS, pH – 3.30, pectin – 0.5-1.0%

Marmalade: Fruits like-Oranges, lemons Preserve: TSS-68%

Jam	Jelly
1. Prepared from fruits pulp.	1. Prepared from cheer fruits pectin extract.
2. Not transparent	2. Transparent
3. Flavours and colours are generally added.	3. Should have the original flavor r of the fruits.

- Preparation of Jam: Preserved SO_2-1000-1500 ppm Fruit part: Sugar - 45:55

  Finished jam should contain 30-50% sugar. Prepared jam is poured hot >85°C into the jars. 40ppm SO_2 may be added for preservation.

  Apple pulp TSS – 15° Brix and acidity – 1.2%

- Preparation of jelly: Guava, apples (Sour and crab) lemons, oranges etc. Extraction of pectin:

  Fruits are cut into pieces of about – 1/8th to 1/4th of an inch in thickness. Time of boiling.

  Apple – 20-25 min.

  Berries & grapes – 5-10 min. Oranges – 45-60 min.

  Guava – 30-35 min.

  Jelly is poured at about 85°C into glass jars.

Determination of pectin content of the extract

1. Alcohol Test
2. Jelmeter Test
3. Making Test Jellies
    - Preparation of marmalade:
        - Generally citrus fruits. Boiled in – 45-60 min. – 103°C
        - 1% acidity as citric acid. Packing – 85°C
        - 40 ppm – SO2 TSS – 65° Brix
    - Preserve, candied and crystallized fruit products:
    - Preserve: Methods of preparation:
        i) Rapid method: Sugar concentration TSS – 68° Brix
        Like soft fruits – strawberries, raspberries
        ii) Slow method: TSS - 60°Brix. Next day - 65°Brix Next day - 70°Brix.

Candied fruits / vegetable

- Bitterness is regarded as a desirable characteristic of marmalades.
- Both ripe and unripe fruits should be used as the farmer yields good flavour and the latter good pectin.
- Protopectin is water-insoluble and when boiled in the presence of acids it is hydrolysed to water-soluble pectin.
- The boiling point of jam and jelly is 105-107°C at the end point.

Canning of fruits and vegetables

- 1804 Nicolas Appert (Father of canning)
  Canning is also called as appetizing or appertization.
- 1817 – Willian underwood introduced coming on a commercial scale in the USA.
- The 1860s: (Levis Pasteur) He introduced the term (Pasteurization) at the temperature of 100°C.

Peeling, Coring

- By hands: All fruits and vegetables
- By machine: Abrasive peelers (potato, carrot turnips)
- By heat: temp. – 40°C about 1 min.

- By key peeling: Peaches, apricots, quinces, carrots sweet potatoes Boiling with soda – NaOH – 1-2% for 30 sec to 2 min.
- 100° salometer reading – 26.5% salt concentration (maximum solubility of salt in water in 26.5%)
- Can filling – 79-82°C temp.
- Canning of fruits – 100°C Vegetable – 115-121°C.
- Fat sour is the spoilage of canned foods due to *Bacillus stearothermophilus*, *Bacillus coagulans* and under processing of cans.

Drying and dehydration

- Drying/dehydration means the process of removal of water.
- Drying: using only sun and wind.
- Dehydration – artificial heat.
- Water activity – 0.50 – 0.65%

Factors affecting the rate of drying

1. Compositions of Raw material:
2. Size, shape, arrangement of food particles.
3. Stacking of Produce:
4. Temp, humidity and velocity of air: Tem. – 50-60°C.
5. Pressure: Boil water – 40-50°C – 28-30 min.
6. Heat transfer to the surface:

Psychometric

- Wet-bulb temp < dry bulb temp

Hygroscopic foods	Non-hygroscopic food
When kept in the open air, they absorb moisture from the surrounding air.	When kept in the open air, they do not readily absorb moisture from the surrounding air.
The partial pressure of water vapour varies with the moisture content of the food.	Partial pressure of water vapour does not vary with the moisture content of food and remain -s constant at different moisture.
Spoilt easily in storage.	Not so
Should be stored in airtight containers agents. packages with added anti-caking agents.	No need to add anti-caking

- Osmotic Dehydration: After lee peeling/blanching/sulphuring / sulphating. Sugar solutions – 60-70° Brix. – hrs tone – a few days.
  - i) Sugar in
  - ii) Water out
  - iii) Minerals, acid, vitamins out. (leaching into the syrup).
- Intermediate Moisture Foods (IME): Moisture – 20-50%

  Ex.: Honey, jam, jelly, cakes, dates, Osmo-dried to.
- Water activity – (d_w)

  $D_w = Pg / Po = N_w / (N_{10} + N_5)$

Conventional drying	Freeze Drying
1. Temp. – 37 to 93°C	1. Temp. - <0°C
2. Pressure usually atmospheric.	2. Reduced pressure 4 mm Hg, 27-133 Pa.
3. Shout drying time <12 hrs	3. Drying time – 12-24 hrs.
4. Principal evaporation of water from food surface.	4. Sublimation of ice to vapour without passing through the liquid phase.
5. Frequently abnormal	5. Usually natural

- Air-blast freezing: temp (-18 to -34°C)
- Cryogenic freezing: temp (-60°C)
- Fruit Beverages: Pure, fruit juice, squash, cordial RTS beverage etc.
- Fermented fruit beverage: Alcoholic fermentation by (Saccharomyces Cerevisiae year) grapes wine, apple cider, berry wine, plum wine, peach wine.
- Fruit squash: fruit pulp / juice – 25% TSS – 40%

  Orange, lemon, pineapple, mango squash etc.
- Fruit juice cordial – juice – 25% TSS – 30%
- Sherbet: Sandal sherbet, Kewa sherbet, Rose sunbet etc.
- Syrup: Brahmi syrup, Rhododendron syrup etc.
- RTS: Fruit juice – 10% TSS – 10%
- Nectar: juice – 20% TSS – 15%
- Fruit juice powder: Orange juice powder Lemon juice powder

Methods of juice clarification

- By sedimentation:
- By use of fining agents:
- Clarification by heating: 82°C temp, 1-2 min. (in pomegranate juice) (lime juice, lemon juice) etc.

- Preservation of fruit juice and pulps:
- Pasteurization: high temp – 100°C long time
- Flash pasteurization: (105.5°C – 1 min) (Orange juice)
- UHT Processing – (130-150°C – 2-3 sec.)
- Pasteurization treatment in milk. LTLT – 62.8° C for 30 min HTST – 71.7° C for 15 sec. UHT – 137.8° C for 2 sec.

Chemical preservative

1. Sodium Benzoate: Test less, water-soluble more effective yeasts then mould, increase CO_2.
2. Sulphur Dioxide: ($K_2 O_2 SO_2$ or $K_2 S_2 O_5$ Potassium Metabisulphite or KMS) KMS is a dry chemical

   Sulphur Dioxide – gaseous SO_2 reduces – O_2
3. Parabens (Para hydroxyl benzoic acids) These are more or less independent of pH.

   Types of Parabens: Butyl, Ethyl, Methyl, Propyl.
4. Nitrates: Most use in meat and its products.

   Types of nitrates – KNO2, KNO3, NaNO2 NaNO3
5. Sorbates: Used in bakery products.
   - Preservation with Sugar: >66% (Specific gravity – 1.330)

     Increase osmotic pressure reduces with activity.
   - Preservation with salt: - 12%
   - Fruit appetizer: juice – 25% TSS – 40° Brix.

Squash	Cordial
1. Contains min 40% TSS	1. Contains min 30% TSS
2. Not a sparkling clear product.	2. Sparkling clear product derived from any suspended material.
3. Prepared from many fruits mango, pineapple litchi, lemon, oranges, plum etc.	3. Mostly prepared from lime juice.

RTS beverage	Nectar
1. 10% TSS	1. 15% TSS
2. 10% fruit part	2. 20% fruit part
3. Artificial coloured added	3. Artificial colour and essence can not be added

- Lime carbonated drink
  Juice – 3%
  Acidity – 0.2%
  TSS – 10-12°b
  CO_2 – 100psi
  Brix acid ratio – 50
- Juice concentrate – 75-90% water
- Clarification of juice: By enzymatic methods (cellulose)
  Enzyme - @0.2-0.3% at 50±2°C for 2 hrs.)
  Crush: 25% fruit part
  55% TSS
  1.0-1.2% acidity
- Syrup: Fruit part – 25%
  TSS – 65%
  Acidity – 1.3%
- Barley water: fruit juice – 25%
  TSS – 30%
  TSS along with 0.25% barley starch.
  Acidity – 1.0%

Alcoholic beverages and vinegar

- Fermentation of sugars by yeast
- Ethyl alcohol – 5-42%
- Wine: Grapes or grapes juice (Saccharomyces cerevisiae)
  Alcohol – 11-14%
  Low – 7%
- Difference between still wines and sparkling wines

Still wines	Sparkling wines
1. Retain no CO_2 *e.g.*, cider	1. Contain a considerable amount of CO_2
Dry wines	**Sweet wines**
Contain no or little unfermented sugars.	Contain unfermented sugars or is added later on.

- Fortified wines: Alcohol – 19-21%
- Table wines: low alcohol and little or no sugar.

- Dessert wines: Fortified sweet wines.

Sherry	Flore yeast
Alcohol	18-21%
Cider	Apple wine
Soft cider	1-5% alcohol
Hard cider	5-8 alcohol
Apple wine	>8% alcohol but can be high as 14%
Perry	Pear wine (could be sweet or dry)
Head	Honey wine
Boukha	Fig wine
Toddy	Coconut wine
Vodka	Prepared from potato
Fanny	Cashew apple wine
Today	Famous wine made in Hungary
Port wines	Grape alcohol – 18%
Nira	Juice of palm tree
Berry wine	Strawberry and blackberry
Sake	Yellow rice beer of Japan – 14-17% alcohol
Sonti	Rice beer of India
Pulque	6% alcohol
Vermouth	Fortified wine – 15-21% alcohol
Brandy	Grape juice (Distillate from alcohol)
Rum	Sugarcane and molasses
Whiskey	Distillate from fermented grains, rye, wheat
Berr	Barley gains
Kirschwasser	Cherry wine
Maraschino	Cherry cordial

- Cider of pasteurization – (62.5°C for 20 min)
- Cider in SO2 @100 ppm

Light wines	Medium wines	Strong wines
1. Alcohol – 7-9%	1. 9-16%	1. 16-21%
2. Not fortified with brandy	2 Sometimes fortified with brandy	2. Are definitely fortified wines

Red wines	White wines
1 Prepared from red/blue coloured grapes high anthocyanin content.	1. Prepared from greenish grapes that have that contain less anthocyanin content.
2. Fruit skin is left in the must for colour	2. Skin is sometimes removed. extraction

- Vinegar
- Cider vinegar: 1.6% apple solids
- More than 50% reducing sugar
- Less than 4% acetic acid at 20°C
- Spirit vinegar: Prepared from ethyl alcohol.
- Wine / grape vinegar: Prepared from grapes / wine.
- Malt vinegar: Malted barley and other cereals.
- Sugarcane vinegar: Sugarcane juice
- Honey vinegar: Honey
- Potato vinegar: Potato starch
- Gain strength: (acetic acid in vinegar)
    - Acetic acid – 4.2%
    - 42% grain strength.
    - Citrus peel is used for the extradition of essential oil.

The enzyme, waste utilization and by brodie

- Enzymes: Enzymes speed up both the deteriorative as well as synthetic reactions.
- PPO – Polyphenol oxidase is the group of enzymes used as an indicator for blanching.
- Postharvest losses – 20-40%

Food processing water utilization and manufacture of by prob

Crop	Waste%	Nature of waste	By-products
Pear	-	Peel core	Animal feed, perry, vinegar
Apple	20-30%	Pomace	Juice, wine, vinegar, pectin, cattle feed
Orange	50%	Peel, seeds, pulp	Essential oil, pectin, cattle feed, peel candy.
Mango	40-60%	Skin or stone	Mango fat, bio, colour
Many peels	12-15%	Peel and pulp	Pectin, cattle feed, alcohol
Pulper waste	5-10%	Fibre	Wine, vinegar, juice
Kernels	15-20%	Holland Kernal	Fat, vitamins, starch
Pineapple	30-60%	Peel, core, Shreds	Juice, wine, syrup, cattle feed, biogas
Banana	20-30%	Peel	Animal feed
Wild apricot	60-70%	Pulp, stone, kernel	Beverages, oil cosmetics products
Aonla	-	Fruit, pulp	Ayurvedic drugs
Pea	60-65%	Pod	Animal feed
Grapes		Stem, pomace	Green of tartar, jelly, chutney
Guava		Cores, seeds, peels	Guava cheese
Passion fruit		Rind, seeds	Pectin oil

- Spoilage of fruits and their products: Spoilage: moisture below – 10% Bacteria grow – 4.8pH
- Lactic acid bacteria: *Lactobacillus plantarum, Lactobacillus brevis, Lactobacillus fermentem, Leuconostoc mesentroids, Pediococcus* sp.
- Acetic acid bacteria: *Acetobacter aceti, Gluconobacter* sp.
- Coliform bacteria: *Escherichia coli, Enterobacter* sp.
- Spore forming bacteria: *Clostridium pasteurianum, Bacillus polymyxa*
- Pathogenic bacteria: *Salmonella* sp, *Shigella* sp.
- Clostridium and Bacillus are spore-forming bacteria.
- Yeasts: Candida spp. *Torulopsis* spp. *Rhodotorula* spp. *Saccharomyces* spp.
- Fungi/moulds: *Alternaria* sp. *Aspergillus* sp. *Colletotrichum sp. Rhizopus* sp. *Sclerotia* sp.

Classification of microorganism based on O_2 recurrent

Type	Molecular O_2 reqd.	Example
Obligate Aerobes	80-100	Bacillus, Pseudomonas
Facultative Anaerobes	50-100	Salmonella, Streptococcus
Facultative Aerobe	20-50	Lactobacillus
Obligate Anarobe	0-20	Clostridium

Classification of Microorganism based on Temp. Tolerance

Type	Temp. requirement (°C)			Example
	Min.	Opt.	Max.	
Psychrophilic	-5	8-10	20	Pseudomonas, Micrococcus
Mesophyllic	5	25-40	55	Feasts, molds, clostridium
Thermophilic	20	50-55	80	*Bacillus stearothermophilus Clostridium thermosacchaslyter.*

Food poisoning

Botulism	**Staphylococcus poisoning**
1. Causal organism (Clostridium botulinum)	1. Causal organism (Staphylococcus aureus)
2. Toxin produced is very potent and a small quantity of it if consumed can cause mortality.	2. Toxin not very potent.
3. Destroyed by heating at 80°C per 30 min.	3. Destroyed at 121.1°C for 10 min.
4. Cause fatigue, digestive disturbances, dizziness, headache etc.	4. Conservation, vomiting abdominal cramping etc.
5. Foods: canned foods, bears sweet corn, asparagus	5. Meat, meat-products milk etc.

Classification of microorganism hard on tolerance to soft and sugar.

1. Halophilic — Salt tolerant microorganisms.
   - Bacteria — Halobacterium
   - Yeast — Torulopsis
   - Moulds — Aspergillus, Penicillium
2. Osmophilic — Sugar loving microorganism
   - Bacteria — Bacillus
   - Yeast — Saccharomyces, Candida
   - Moulds — Aspergillus

Food spoilage

Marketing and export of fresh and processed products

- The market is a Latin word – meaning of (Merchandise) (Banana) share in consumer's rupee is around 31-35%. Transportation:

Trucks	500km
Rail	1000km
Jeep/Lamper	30-35km

- Agricultural and processed food products export development authority (APEDA) – 15 Dec 1985.

  But now processed food export promotion council (PFEPC)

  Planning of fruit and vegetable processing industry
- Novel technologies in food preservation:
- Extrusion: "If the food is heated, the process is called as extension cooking".
- Extrusion is a size enlargement process.
- Pressures of 1500-70000kPa and temperatures of 52°C-199°C are generated during extrusion.

Irradiation

- Thermo radiation: 210 k rads - 27°c 157 k rads - 60°c
- Delay of fruit ripening: (0.25-0.75 kgy.)
- Sterilization of herbs and spices - (8-10 kgy)
- Sterilization of packaging material – (10-25 kgy)
- Sterilization packaged meat, poultry product – (25-70 kgy)

- Sterilization of hospital diets – (25-70 kgy)
- 1gy = 1j of energy absorbed per kilogram of food.
- 1 Rad = $10^{-2}JKg^{-1}$ 1gy = 100 rads.
- Only Y-rays from cobalt-60 and caesium – 137.

  The order of sensitivity of various vitamins in decreasing order is: thiamin > ascorbic acid > pyridoxine > riboflavin > folic acid > cobalamin > mioclinic acid.

  Vitamins D and K are practically unaffected while Vitamins A and E undergo some losses.

Ohmic heating

- Energy type and wavelength

Cosmic rays	Very shout
Alpha, beta, gamma rays	<100 nm
X rays	100-150 nm
UV	13.6-400 nm
Visible	400-800 nm
Microwaves	30 cm
Dielectric	25m
Radio waves	30m-3 km.

- Microwaves ovens are thus - 40% efficient as compared to 14% for standard electric ovens and 7% for gas ovens.

  Microwaves are generated by magnetron at (2540 MHz) or sometimes 896 MHz in Europe and 915 MHz in the USA.
- Pulsed electric fields – started 1920's-1930's First use in milk in the USA

  TMP of cell = 1 volt

  CEFS – Critical electric field strength – 15kv/cm

  Lactobacillus brevis and *Escherichia coli*-13kv/cm. and 16kv/cm respectively.
- Magnetic field:

  Magnetic field intensity is between – 5-50 Tesla while the shout 25 µsec up to few minutes and frequency of 5-500 kHz are required for inactivation of microorganisms.
- High-pressure crossing: 350MPa for 30 min 400 MPa-5mi cause about 10 folds reduction in vegetative cells of bacteria, yeasts or moulds.

  Little chance of any toxicity.

- Ultrasounds: Frequencies above 16KHz that cannot be detected by a Hammer. Low intensity - <1W Cm-2

  High intensity – 10-1000 W Cm-2 Higher intensity – 2.5 MHz
- MTS – (mano-thermo-sonication) heat and ultrasound treatment
- LIEA- linear induction electron accelerator.

Minimal processing and hurdle technology

Some active compounds

Compounds	Plant in which ford	Properties	Use
Butylphthalide	Celery	Provides distinctive taste and small	Protection against cancer, high blood pressure and high cholesterol levels.
Calcium pectate	Fruits and mega-apple, carrots, onion, cabbages	Responsible for crispness in fruits and vegetables.	Potent cholesterol lowering properties.
Catechin hydrate coumarin	Tea, Tonka bean lavender, sweet clover grass licorice strawberry apricots, cherries	A tannin derivative that gives tea its astringency. Blood thing agreed that set as a rat passion.	Cancer fighting properties have been attributed to tea anti fungicide and anti-tumour activities.
Ellagic acid	S+B and Resb. Highest qty. Block b. Cann b. Walnut and pecans.	A natural phytochemical pesticide in many fruits plants.	This phytochemical fights Cancer in humans.
Lycopene	Tomato, watermelon, grapefruit, guava, worship and red chilis.	Responsible for the red-orange and yellow colour in fruit and vegetable.	Protects against heart disease, certain cancers and a multitude of other disorders.

- GM (Genetically modified foods) – Soybean, corn, canola and cottonseed oil.
  - Euro – Retailer produces working group – (EUREP-1997).
  - Prevention of food adulteration act – 1954 (PFA) vegetable product order – VPO – 1967.
  - Meat and food products order – 1973 (MMPO)
  - Agri-produce grading and marketing act – 1937. Bureau of Indian Standard BIS, Export (Quality control and inspection act – 1963.
- Codex covers 234 foods, 32 functional groups (additive) 36 flavouring agents, 6 labelling standards, 49 condos of practices and 41 guidelines.

- BIS has about 700 Indian Standards for use in the area of agricultural produce and value-added products.
- HACCP (Hazard Analysis and Critical Control point)

  Seven Principles / Components of HACCP

  i) Hazard Analysis

  ii) Critical control point

  iii) Critical control for each CCP

  iv) Monitoring

  v) Corrective actions

  vi) Record keeping

  vii) Validation and verification

3

Plant Propagation in Fruit Crops

- Grape, fig, pomegranate, olive – hardwood cutting pineapple, date palm – suckers (blueberry) raspberry.
- In general, the self-pollinated plants, which are considered as homogeneous, are propagated by seeds.
- Runners : Strawberry
- Offsets : Pineapple, Banana
- Hardwood cutting : Grape, fig, pomegranate, mulberry, kiwi fruit, olive, quince.
- Semi-hardwood cutting : Mango, guava, jackfruit, lemon.
- Soft wood cutting : Junips, hollyhock etc.
- Leaf bud cutting : Blackberry, lemon.
- Root cutting : Apple, pear, cherry, fig, blackberry.
- Tip layering : Trailing blackberry, raspberry.
- Simple layering : Lemon, grape.
- Tooling/mound/layering : Apple rootstock, guava, mango, litchi, gooseberry.
- Serpentine layering : American grape.
- French layering : Apple rootstock, cherry, plum.
- Craft grafting : Avocado, apple, pear, plum.
- Side grafting : Mango, avocado.
- Inarching : Mango, sapota, guava, litchi.
- Veneer grafting : Mango.
- Stone grafting/epically : Mango, walnut.
- Top working : Mango, ber, cashew nut, mulberry.
- Softwood grafting : Cashew nut.
- T-budding : Citrus, plum, peach, cherry, ber.

- Patch budding : Walnut, pecan nut, Indian rubber.
- Chip budding : Mango, grape etc.
- Ring budding : Peach, plum, ber, mulberry.
- Shoot tip culture : Citrus, papaya, grape, strawberry.
- Embryo culture : Citrus, rhododendron.
- Single-cell culture : Carrot, suflow.
- Embryo rescue : Grape.

Plant structure

- Peach – In leaf axils is called – as (blind bud).
- Pecans and walnut, 2-4 buds development at one bud.
- Plants which grow from adventitious buds on the roots are called suckers.
- Some are broad-leaved evergreen: Citrus, mango, litchi, avocado.
- Narrow-leaved evergreen – (coniferous trees)
- Deciduous plants: apple, pears, grape, apricots, pecans.
- Monoecious: Walnut, pecan nut, hazelnut, Tung coconut, cucurbits.
- Dioecious: Papaya, date palm, pistachio nut muscadine grape pointed and snake gourds asparagus.
- Spike inflorescence: Banana
- Catkin: is a spike pecan nut, oak, mulberry, dates and walnut.
- Raceme: Pear
- Corymb: Cherry, pear, candytuft.
- Fascicle Inf.: Phlox and apple.
- Panicle: Mango, grape, litchi.
- Self-pollination (autogamy)
- Cross-pollination (autogamy) (gait gamy)
- Dischongamy: Walnut, pecan nut, macadamia nut.
- Self sterility: Almond, some fruit, onion, marigold.
- Incompatibility: Mango, ber, loquat, apple.
- Wind pollination: Walnut, pecan, nuts, coconut, dates.
- Insect pollination: Apple, mango, peach, fig, coconut.
- Birds: Pineapple and banana.

- Water: Water lily, water chestnut.
- Seedless fruits: Banana, pineapple.
- When plants set and mature fruits without pollination, it is called as vegetative parthenocarpy. Banana, Japanese person, fig, orange, grapefruit.
- Monocarpic: One crop of seed and then dies.
- Polycarpic: One crop of seed and then does not die.
- Media for propagation: 5.5-6.5 pH.
- Sand – 40%

  Silt – 40% Germination media

  Clay – 20%
- Sand size – 0.05 – 2 mm
- Vermicompost can be divided into four groups:
  1. Grade has particles from – 5-8 mm
  2. Grade has particles from – 2-3 mm
  3. Grade has particles from – 1-2 mm
  4. Grade has particles from – 0.75-1 mm
- Peat Moss peat: Sphagnum or hypum, highly acidic pH-3.2-4.5, nitrogen little or no Pand K.
- Reed sedge peat: Grasses reeds, ridges and swamp etc. colour reddish and blackish pH 4.0-7.5.
- Peat humus: Colour brown and black (very popular propagations peat).
- Sphagnum moss: pH – 3.5 to 4.0 use in layering (Cgoottee) rooting media.
- Pumice: Varying amount of iron, calcium, magnesium and sodium.
- Perlite: Volcanic origin – 760°C, pH – 6-8, size – 3-8 mm.
- Leaf mould

Propagating, structure and equipment

1. Greenhouse

- Low-cost greenhouse: Polythene thinners 0.10-0.15mm resistant to ultraviolet rays.
- Net house: Use in tropical areas, artificial cooling, gamy cloths.
- Bottom heat box: Two chambers
  - Height – 70 cm.

    - Width – 40 cm.
    - Outer chambers
- Inner chambers:
    - Height – 68 cm,
    - Width – 44 cm.
    - Temp.: 30±20 °C

Sexual propagation

Seed Dormancy

- KNO3 or GA3 temp – 20 °C
- ABA can inhibit germination in non-dormant seeds.
- GA and Cytokinin application is required to overcome seed dormancy.
- Dormancy, used in combinations with kinetin, CO2 or light treatments.
- To overcome dormancy: KNO3 – 0.2%

Softening seed coat

Scarification

- Acid scarification: dry seeds: sulphuric acid 1:2, 15m – 6 hrs.
- Hot water scarification: 77 to 100°C, 12-24 hrs
- Water: seed – (1:5)
- Warm moist scarification:
- Coldwater: apple – days
- Mulberry – 4 days
- Stratification: Chilling temp – (5-10°C)
- Refrigerated stratification: 12-24 hrs. temp (0-10°C)
- Seed: Peat – (1:3)
- Some seeds: (1-4 months) like, apple, pear, plum.

Outdoor stratification

- Pre chilling: Kept at a temp of 5-10 °C for 5-7 days sowing.
- Pre drying: Kept at a temp of 37-40 °C for 5-7 days sowing.
- Hormonal treatments: GA3 commercially uses for breaking seed dormancy.
- GA3 - 200-500 ppm
- Cytokinin (Kinetine and BA) soaking seeds in a 100 ppm of the solution of Kinatine 3-5 min.

- Chemical treatment: Thoreau – 0.5-3% solution for 3-5 minutes.
- Seed priming: Polyethylyol (PEG) 20-30% solution, incubated at 15-20°C for 7-21 days.

Seed germination

- Radicle – Root
- Plumule – shoot
- Seed germination

Stages of germination

1. Activation stage
    a) Water absorption
    b) Synthesis and activation of enzymes: (Hydrolytic enzyme) Convert complex food material into a simpler form.
    c) Cell elongation:
2. Translocation stage: Proteins – amino acids Starch – Sugar
3. Seedling growth stage:
    - The section of seedling stem above the cotyledons is called as epicotyls and below the cotyledons hypocotyls.
    - In the first type, the hypocotyls elongate and raise the cotyledons above the ground surface: it is called as (epigeous or epigeal germination) e.g beams.
    - In the other type, the epicotyls elongate and hypocotyls do not repair the cotyledons above ground. It is called Hypogeous or Hypogeal germination mango, contend apple, lotus.
    - Germination is waterlogged soil also because they have low O_2 requirement.

Seed viability

- Some seeds have long-life if stored at low temp (0-4°C) and low humidity, whereas seeds of some species litchi, mango, citrus, cocoa, walnut are short life for germination.

    a) Orthodox seeds: Ber, custard apple, date palm, fig, grape, guava, mulberry, papaya, passion fruit, peach, pineapple, plum, phalsa, pomegranate are orthodox seeds.

b) Recalcitrant seeds: Avocado, Barbados cherry, carambola, breadfruit, durian, jackfruit, litchi, mango, mangosteen, rambutan, walnut and citrus, Jamun.

- The longevity and storage condition for seeds of different fruit and plantation crops.

Fruit crop	Longevity	The ideal environment for germinations	Remarks
Arecanut	Very less	Cannot be dried	
Avocado	15 month	5°C, 90% R.H.	
Barbados cherry	-	can't be dried	
Breadfruit	very less	can't be dried	
Carambola	very less	can't be dried	
Grapefruit	12 months	5°C, 85% R.H.	
Jackfruit	-	5°C, 85% R.H.	
Lemon	8 months	2%, 88% R.H.	Treat fungicide
Lime	6 months	2°C, 88% R.H.	Treat fungicide
Litchi	3 weeks (85% R H.)	3°C, High R.H.	Proper moisture
Loquat	6 months 92%	5°C, high R.H.	Semi sealed con.
Macadamia nut	6 months 50%	Ambient temp and R.H.	
Mandarin	4 year 88%	7°C, 58% water	
Mango	80 days	Room temp 50% R.H.	Temp. – 3°C
Mangosteen	8 weeks	21-29°C	
Persimmon	Can't be dried		
Peemmelo	8 months	2°C, 88% R.H.	Treat with fungicide
Rough lemon	16 months (46%)	5°C, 46% water	Apple moisture, grey land
Sapota	Can't be dried		
Gour oranges	11 months	5°C, 88% R.H.	
Cinnamon	1 month	Ambient top	
Cocoa	4 months (22%)	10°C	24% water in seeds
Coffee	47 weeks (22)	warm humid	0°C or 10% water
Tea	10 months (50%)	0°C, 100% R.H.	

- Factors affecting seeds viability: Mechanical injuries:
    - Environmental conditions:
    - Genetical factors:
- Indian lotus has been reported to remain viable for 100 years. Storage condition:
- Seeds moisture 5-14 %
- Methods of testing seed viability:

- Germination test:
- 2,3,5 - Triphony tetrazolium chloride known as (tetrazolium salt) pH- 6-7, the chemical is highly sensitive to light.
- Excised embryo test: Room temp-18-20°C R.H. Normal light intensity.
- X-ray tests: In x-ray test the seeds are first soaked water for 14-16 hours and then placed in a concentrated solution of (Barium Chloride) for 1-2 hours water, the excess barium chloride is removed by washing seeds in running water and then photographed by soft x-ray.

Seed Production Testing and certification

- Seed Act – 1966, Govt. of India.
- Similarly, the light of 750 to 1250 Klux should be maintained in the germination for proper germination.
- Breeder's seed: Golden yellow
- Foundation seed: White
- Certified seed: Blue
  - Registered seed: Green
  - Nursery establishment: pH – 5.5-6.5
- Seed sowing: June-July – Mango, Kagziline, Jack. Feb-Mar – Guava, ber, aonla.
- In sites sowing: Walnut, pecan, jack, ber.
- In rocky soils of Orissa and Gujarat, in sites sowing of mango seeds and HDP like Amrapali is practised .
- Zinc – 20-25 kg/ha or its foliar application – 0.5%
- Iron – 40-50 kg or its foliar application – 0.5%
- Manganese – 20-25 kg/ha or its foliar application – 0.2-0.4%
- Copper– 10-12 kg/ha or its foliar application – 0.5%
- Boron – 15-20 kg/ha or its foliar application – 0.2-0.5%
- Mo – 1-2 kg/ha or its foliar application – 0.1%
- Damping-off: Pythium, Phytophthora, Rhizoctonia, Botrytis cinerea.
- Scales – Roger or Metarystox – 0.05%
- Thrips – Metasystox – 0.05%
- Mealybugs – Dimethoate – 0.05%

- Whiteflies – Phosphamidon – 0.02%
- Mites Dicofol – 0.05%, Sulphur – 0.2%
- Leaf eating caterpillars – Sewin / Carbary – 0.1%, Neem oil- 1%.
- Cutworms – Malathion – 0.1%, Quinalphos – 0.05%.
- Leaf miners – Metasystox or roger – 0.05%
- Snails and slugs – common salt – 2%
- Damping-off – Thiram or agrosam @2g/kg/h.
- Powdery mildew – Prophylactic spray of kasathane or calixin or sulphur – 0.2%
- Leaf spots – Dithane (-) z-78 or bavistin – 0.2%.
- Blights – Z-78 or bavistin – 0.2%
- Dieback – Benlate – 0.2%

Marketing of Nursery Plants

- In general, the polybag of 30 cm × 20 cm size is used for raising fruit plants.
- The pH of the solution can be adjusted with the help of 0.1 NHCL or 0.1 NKOH.

Routine Nursery Practices Vegetative Propagation

- This capacity of a plant cell is called is totipotency.
- The term totipotency was first coined by a physiology Haberlandt in 1902.
- Bartlett pear originates during – 1770 in England.
- Delicious apple in Peru during – 1870.
- Shoot tip culture – Virus free, like apple, peach, grape, strawberry, Chinese gooseberry, citrus, pineapple.
- Thermotherapy: temp – 38 to 40°C (control pathogens)
- Chemotherapy: Chemical (control pathogens)
- Mutations: First coined by De varies in 1905.

Mutations

- Grapefruit: Pink fleshed (instance) White fleshed (bud spot)
- Orange: Washington Navel – (bud mutant)
- Mango: Dashehari – 51 is a regular bearing mutant.

- Anatomical basis of rooting of cutting it was first time given by French dendrologist Duhamel Du Monceau in 1758.
- Auxins: Indole-3 acetic acid (IAA) was identified as the fruit naturally occurring compound, during 1934.
- Use of synthetic auxin in 1935.
- Cytokinins: Synthetic chemicals: Kinetin, Sentin 6-Benzyl adenine. Gibberellin: First in Japan – 1939.
- Ethylene: 10 ppm causes root formation in stem and root cutting and inhibit 100 ppm and above.
- Thiamine (Vit. B1), pyridoxine (Vit. B6) nianic and biotin B-complex Vit. And Vit K or Hall are known to stimulate the rooting process.
- Boiullenne and Went were the first to name the root forming factor of leaf, cotyledon and bud as rhizoplane in 1933.
- Cutting for – low nitrogen and high carbohydrate.
- Hardwood cutting of Bartlet pear does not root easily as compared to old hume cultivar.
- In hardwood cuttings, more roots are developed in the basal portions of the shoot than the cutting has taken flonicamid or top of the shoot.
- Viruses not only reduce rooting % but also the root number.
- Best roots in winter – Hardwood cutting Best roots in spring – Soft wood cutting.
- Treat cuttings with Mn1B and P but better results were obtained with potassium permanganate.
- Destroyed by bacteria (Acetobacter destroys) IAA but not IBA, NAA or 2,4-D.
- IAA is highly sensitive to light and is destroyed by strong sunlight through NAA and 2,4-D is stable.
- Bron also plays an important role in the rooting process.
- The combination of nutrients (N1B) with auxin (IBA) is the most effective treatment for root initiations and development.
- The treatment of cutting caption and benomyl gives the best results.
- Girdling in girdling a ring of bark – 2.5 cm to 3.0 cm. removed from the base of the shoot.
- A day temp 21-24°C and night temp 13-15°C is optimum for rooting of cutting.

- The orange-red light of the spectrum seems to favour rutting of cutting then the blue region.
- Further, red light (680 nm) has more inhibitory influence on rooting than the blue or far-red light.
- Oxygen (1-10 ppm) in the medium is essential for root initiation process.
- Hardwood cutting: length: 10 cm – 45 cm, diameter- 0.5 cm - 2.5 cm.
- Semi-hardwood cutting: Used in an evergreen plant like mango, guava, lemon, length: 7 cm to 20 cm, (5000 ppm IBA).
- Softwood cuttings: Length – the 5.0 cm - 7.5 cm best time for preparing softwood cutting in late summer.
- Herbaceous cutting: Length 7 cm - 15cm requiring high R.H. eg. Pineapple.
- Root cutting: Length- 10 cm - 15 cm, thickness – 1cm, time December.
- Root cutting: horizontal planting – blackberry, raspberry, kiwi fruit, breadfruit fig, mulberry, apple, pear, peach and persimmon.
- Leaf cuttings: Lemon
- Bud cuttings: 2.5 cm in length.
- Eye cuttings: Vines, 4 cm – 5 cm long having on the eye or a.
- A bottom heat of 24°C is required for rooting.

Propagation by grafting

Grafting fro temp. – 30°C

- New cambial calls after 2-3 weeks of grafting.
- Top working of apple and peas with any simple technique do well but difficulties appear while top working the stone fruits.
- Top working of peaches with compatible species of plum or almonds is more successful than working with peaches alone.
- Muscadine grape (*Vitis rotundifolia*) and mango with simple budding and grafting techniques for which approach grafting or often used and to have better success.
- In apple, little or no callus is formed below 0°C and 40 °C optimum being 24- 27 °C for walnut it is 25-30 °C.
- It is highly successful propagation method for mango in Konkan region but not in North India.
- Grafting is generally confined to dicotyledonous plants & some extent in gymnosperms because they have a vascular cambium layer between xylem & phloem.

- Monoctyledence plant doesn't have the cambium tissue and so grafting is not successful in them except some grasses and tropical orchid vanilla .
- Dashehari mango tree can be grafted successfully on the same Dashehari tree.
- Grafting between species of a genus: For example, grafting between most species of citrus is highly successful and used commercially.
- The almond, apricot, Japanese and everbearing plum are grafted commercially on peach which is a different spp.
- But almond and apricot; both in the same genes, can't be inter-grafted successfully.
- Beauty cultivar of Japanese plum makes successful union when grafted on almond but Sentosa does not.
- Japanese plum cultivars can successfully be grafted on European plum.

Craft incompatibility

- Some pear cultivars are successfully grafted on quince rootstock whereas the others maybe soon.
- The quince on pear rootstock is always a failure.
- Plum grows well on peach RS bud peach graft age on plum is always a failure.
- Kinnow mandarin budded on Troyer citrange usually develops an outgrowth above the bud union.
- Allahabad Safeda scion of guava usually develops on outgrowth above the grafted on dwarfing Aneuploid No. 82.
    1. Localized incompatibility: Reactions are quite common in apple, when grafted on pear and plum on cherry.

       The reaction is that of Bartlett pear grafted on quince rootstock.
    2. Tran located incompatibility:

       Phloem degeneration takes place resulting in the formations of brown line and narcotic area in the bark.

       Ex.: That Halers Parly peach grafted on Myrobalan B Plum RS.
- Almond is highly compatible with Mariana 2624.
- Pear cultivars are grafted on to quince, a gynogenetic compound (prunasin) found in quince, not in pear.
- The pear tissues breakdown Premarin with hydrocyanic acid as one of the products.

- Jonathan apple is grafted on to EM-9 rootstock, the scion develops molybdenum (Mo) deficiency but Jonathan on other rootstock does not show such deficiency symptoms.
- P.K. and Mg in peach has been reported if grafted on to Myrobalan B plum rootstock, resulting in incompatibility symptoms.
- Presence of latent viruses and many plasma-like pathogens may fail graft unions.
- Incompatibility reactions in citrus are primarily due to infections by viral durians like Tristeza, porosity and xyloporosis etc.
- Inarching grafting – adjuvant grafting
- Notch grafting – It is also called inlaying. It is particularly useful in humid tropical areas.
- Defoliation of the shoots is not required under humid conditions.
- The best time for veneer grafting in North India in March – April and July – August. (Ambient temp – 25 ± 35°C)
- Cleft or wedge grafting: Also used for rejuvenating the old orchards of tempered fruits by top working.
- Root grafting: Apple, pear, plums, filberts and wisteria etc temp. – 35-40 °F.
- Double working in apple and pear.
- Greenwood grafting: Grape.
- Nurse – Seed grafting: Chestnut, avocado, canella pecans and walnuts.
- Cutting-grafts: Citrus and donal rootstock.
- Softwood grafting: Commercial in mango, sapota, tamarind and cashew nut in western India. The technique is applied to 60-70 days old rootstock. (poly – 100 gouge)
- Epicotyls grafting: Poly – 150 gauge in mango.
- Micrografting: Citrus, apple, plum, first developed by Spanish scientist, Nauarroand his worker.
- Mazzard (*Prunus avium*) vigorous rootstock of sweet cherry.
- Sour orange rootstock produces thin skimmed juicy and excellent quality fruits.
- Rough lemon r/s produces thick, courses skimmed fruits of inferior quality.
- The bitterness of the grapefruits disappears if budded on trifoliate orange RS but it never disappears on Rough lemon rootstock.

- Troyer citrange rootstock usually reduces fruits size in Kinnow but Karna Khatta and Sohsarkar improve it.
- Fruits on Troyer citrange are thin-skinned with bright orange colour, good TSS, compared to those obtained from Karna Khatta and Sohsarkar rootstock.
- Old home pear is resistant to pear decline.
- Mahaleb rootstock of cherry imports resistant against best skin virus.
- Apple rootstock to resistant winter injury EM-9, 17, Alnarp-2.
- EM-9 is resistant to early winter frost.
- Trifoliate orange is considered as the hardiest rootstock to low temp followed by sour orange as sweet orange.
- Jatti Khatti citrus rootstock performs well in semi-arid zones Punjab and Rajasthan.
- Karna Khatta is suitable in Indo-Gangetic plains of UP but not the others.
- Myrobalan plum is highly suitable rootstock in regions of excessive soil moisture. Whereas almond is most susceptible to wet conditions.

Budding: Theory and Principles

- Fall budding: time – mid-June to mid-September.
- Spring budding: The buds must be dormant and the stock must be in active growth. Spring budding has commonly flowed in citrus.
- June budding: New shoot, which developed in the spring. June budding is commonly in-store fruits like peach, plum, apricot cherry, nectarines and aonla.
- Bud union: (i) Pre-callus stage, (ii) Callus formation, (iii) Formation of cambium bridge, (iv Healing of the bud union.
- Pre-callus stage: 5 to 8 days after budding.
- Callus formation: After 5th to 8th day, usually living cells of recently formed xylem and phloem plays an active role in the callus formation.
- Formation of the cambial bridge: The parenchymatous cell of the callus forms a cambial bridge within 12-15 days after budding.
- Healing of the bud union: Many nqew xylem and phloem tissue after 6-8 months of budding within 12-15 days after budding.
- Shield or T budding: Rootstock 25 cm - 35 cm hight and 2 cm - 2.5 cm on thick.

- Chip budding: Grape, apple, pear are propagated by these methods, besides, grape in the dormant stage is also budded on nematode of phylloxera tolerant rootstock of the method.
- 'I' budding is more successful in plants having thicker bank than that of bud stick.
- Skin budding: It is very useful in citrus.
- Micro budding: Citrus in Australia.

Rejuvenation of old orchards

- Top working: Mango, cashew, ber etc.
- Heading back: Cut the breaches at the place where the plants are 6 to 9 inches in diameter.
- Best time and temperature for budding and grafting: temp-90° to 100°F Spring in plains and the summer months in the hills.

Apomixes

- Recurrent apomixes: Raspberry, Mallus sp. Onion Parthenium. (diploid egg)
- Non-recurrent, apomixes: (Haploid egg) Eg. *Solanum nigrum* and ilium
- Adventitious embryonic: (nucellar embryonic) e.g. citrus.
- Vegetative apomixes: eg. Allium, Agaue, Poa and Dioscorea.
- Polyembryony: eg. Citrus, mango, Jamun.
- Apomictic seedling: Citrus, apple (*Malus toringoides*, *Malus sikkimensis*)

Propagation by layering

- Tip layering: Blackberries, raspberries and dewberries.
- Tranch layering: Apple-like M16 and M25 and walnut.
- Mound layering: Apple, guava, quince, plum, cherry, hazelnut, peanut, mango, jackfruit & litchi.
- The weight of a tuber price should vary between 28 to 56g.
- Bulbs are usually produced by monocotyledonous plants.
- Non-tunicate bulbs: garlic, lilies.
- Bulbs, the seals with NAA (IBA) is useful for stimulating bulblet formation.
- Most of the rhizomes are monocotyledonous but dicot blueberry.
- Pineapple is commercially propagated through slips.Use of chemicals in plants propagation

- Soil treatments: Formaldehyde (2%) solution
  - Chloropicrin, methyl bromide, captan and benomyl.
- Seed treatments: Ethyl alcohol %, calcium hypochlorite %, Mercuric chloride 0.1% seeds, usually from 5-10 minutes. organomercurials: Aerosan-6N
  - Non-organic-mercurial: Brassicol, captan thiram, Dithane M.45.
- Dormancy: GA-500-1000ppm
  - Kinotin-100ppm
  - Thiourea-3%
  - Sodium hypochlorite
  - Calcium hypochlorite
  - Seed viability: through tetrazolium test.

Role of plastics in plant propagation

- Polyethene: A greek words.
  - Poly – Many
  - Ethylene- from crude oil.
- This was discovered by the United Kingdom in 1933.
- The sheet film, which is invaluable in gardening, was manufactured in 1938 in the United Kingdom.
- Polytunnels: 200-300 gauge.
- Polythene use in budding and grafting: 150 gauge thickness.
- The earthen pots of polythene: 200-500 gauge.

Micropropagation

- Tissue culture: German plant physiologist, Haberlandt 1902.
- Tissue culture systems produce secondary metabolites, which are used by many pharmaceutical industries.
- Explant: Treated with different sterilizing agents mercuric chloride, hydrogen peroxide sodium hypochlorite, silver nitrate, bromine water or ethanol before the culture.
- Culture media sterilizing: 121°C for 15 min.
- MS media: Murashige and Skoog – 1962.
- WPM Media: (Woody plant medium – 1963).

- Nitsch & Nitsch 1956 medium.
- Media pH – 5.7 to 7.8
- Blue light favours shoot initiation.
- Temp in fruits: 27±2°C
- Temp in bulbs and carnation 25±2°C
- Anther culture: Haploid plants were discovered by Giha & Maheshawari 1959.
- Among fruit crops, strawberry was the first fruit to be propagated commercially through micropropagation.
- Biotechnology of fruit and nuts crops

4

Physiological Disorders in Fruit Crops

Aonla

1. Fruit necrosis

- Fruit necrosis is especially a problem in 'Francis'
- Boron deficiency / Borax at 0.6%, September-October.

2. Fruit drop

i) First wave: 70% lack of pollination

ii) Second wave: Dormancy break

iii) Third-wave:

Dry spell, number of developing fruits. Imbalance of hormones and improper nutrition.

Fluctuation in temperature, variety and age of the tree.

Apple

1. Scald

Granny Smith, Rome Beauty, Delicious, Winesap and Yellow Newton varieties are extremely susceptible whereas Gala and Fiji are moderately susceptible.

Symptoms: Greener surface / Brown patches.

Causes

1. Hot and dry weather before harvest
2. Immature fruit at harvest.
3. High nitrogen and low calcium concentration in the fruit.
4. Inadequate ventilation in storage rooms or packaging boxes. **Control:** Antioxidant immediately after harvest CaCl2 (2-3%)

2. Bitter pit

The fruits of Northern Spy, Golden Delicious, Yellow Newton and Gravenstein are most susceptible.

Causes

1. Low level of calcium
2. High level of nitrogen
3. Excessive shedding and heavy pruning
4. Early and over thinning of fruits

Control: Calcium sprays

3. Internal Browning (Brown heart)

- High CO2 concentrations in storage
- Large and over mature fruits
- The incidence of the disorder is less at 1°C in cold storage

4. Cork Spot

1. Low calcium in the fruit
2. High level of Nitrogen
3. Unfavourable soil conditions
4. Excessive tree vigour **Control:** CaCl2 – (0.5%).

5. Water Core

1. Large fruit and high leaf to fruit ratio
2. High fruit nitrogen and boron

6. Sunburn

- Due to the intense heat of the sun
- Water stress
- Granny Smith more sensitive to sunburn

7. Russeting

- Improper training and pruning
- Excessive nitrogen application

8. Jonathan Spot

- First described on Jonathan apple and is perhaps most common on this variety
- Brown to black spots
- Jonathan spot is common after a dry spell

9. Fruit drop

- o Hormonal imbalance, particularly Auxin
- o 10 ppm NAA 20-25 days before harvest

10. Alternate bearing / Biennial bearing

Flowering promoters such as Ethephon may help boost return bloom after an 'On' year & floral inhibitor's such as (GA3) may reduce bloom in the season following the 'Off' year.

Apricot

1. **Pit Burn**: Higher temperature and longer duration
2. **Gel breakdown or chilling injury**: Stored at low temperature between 2.2-7.6°C

Avocado

1. **Leaf burn:** accumulation of sodium and chloride in irrigation water

Bael

1. **Fruit cracking:** Sudden changes in weather conditions such as temperature and humidity. **Control:** Borax – 0.1%
2. **Fruit drop:** Imbalance of soil moisture, improper nutrition and hormonal imbalance. Spraying growth regulators like 2, 4-D, 2,4,5-T & GA3.
3. Storage of fruits below 9°C

Banana

1 **Neer Vazhar:** Mycoplasma like organisms (MLO's). **Control:** NAA

2 **Kotta Vaghai:** Virus

3 **Derain:** Due to rotting of pedicels which results in droping of ripe fruit from the bunch.

To control degrading, excessive application of N fertilizers should be avoided.

4 **Peel splitting:** Borax – 0.8%

5 **Chilling injury:** extremely low temperatures at the time of storage

6 **Choke throat:** Choke throat is seasonal as it is usually worst during the winter and early spring following cold weather.

Higher Nitrogen rates are thought to be beneficial

Ber

- **Fruit drop:** The initial fruit set is very high. 20-30ppm NAA during the second week of October and November.
- **Chilling injury:** Low temperatures at the time of storage

Breadfruit (*Artocarpus altilis)*

1. **Chilling injury:** Storage temperature below 12°C

Carambola (*Averrhoa carambola*)

1. Chilling injury: Storage at low temperature

Cashew nut

1. **Yellow leaf spot:** low soil pH – 4.5-5.0 due to Molybdenum Deficiency Spray with 0.3% Ammonium Molybdate. June and Sep.
2. **Little leaf of cashew:** Due to deficiency of zinc. Zinc sulphate – 0.5%

Cherry

1. **Fruit cracking:** CaCl2 @300-500g/100 lt. of water. GA3 2000ppm 3 weeks before harvesting.
2. **Surface pitting and bruising:** Most common in Sweet Cherries

Citrus

1. **Fruit cracking or splitting:** Hot winds or deficiency of boron and calcium are responsible for this disorder. NAA – 100ppm, Borax-0.8%.
2. **Fruit drop:** 2,4-D-20ppm
3. **Citrus decline:** Excess of boron and CaCo3, deficiency of Zn and Mg.
4. **Granulation:** First reported in California in 1934. 2,4-D-16ppm. Mixture of– Zn, Cu and K each (@ 0.25% should be applied during August-September
5. **Frenching / Little leaf:** Zn deficiency, Zinc sulphate – 0.5%.
6. Oil spotting (Oleocellosis):

7. **Chilling injury:** Lower temperature - 10°C at storage
8. **Frost injury:** exposure to temperature below – 2.2°C for more than 4 hours.
9. **Albedo breakdown or creasing:** Creasing is caused by excessive loss of cohesion between albedo cells; the white layer under the skin which gets stressed by the expansion of the pulp. **Control**: GA3 and Calcium
10. **Puffing:** over maturity, early picking
11. **Sunburn:** Direct sunlight or hot dry winds.
12. **Watermark/ waterburn:** This occurs during wet weather when the bottom of the fruit is wet for a prolonged period. **Control**: GA3

Coconut

1. Barren nuts
   - Poor pollination
   - Deficiency of minerals such as K and B.

     1kg Muriate of potash and 200g Borax per plant
2. **Button shedding and premature nut fall:** Lack of pollination: 2,4-D @ 30ppm.
3. **Crown choking:** Boron deficiency 200g of Borax per plant.

Custard Apple

1. **Stone fruit:** Nutritional deficiencies
2. **Fruit cracking:** Sudden changes in weather conditions such as temperature and humidity
3. **Decline:** Water stagnation / Water logging

Datepalm

1. **Blacknose:** High humidity / Rainfall / over thinning.
2. **White-nose:** Dry ring at the calyx end of ripening fruit. Prolonged dry winds in the early rutab stage cause rapid maturation and desiccation of the fruit.
3. **Crosscuts:** (V-cuts) in the tissue Khadrawy variety is most susceptible to this disorder.
4. **Black scald:** High temperature is considered the main cause behind this disorder.

5. **Bastard offshoot:** Due to reduction in growth and imbalance of growth regulators.
6. **Darkening:** High moisture and high temperature
7. **Skin separation or puffiness:** High temperature and humidity

Durian (*Durio zibethinus* L.)

King of all fruits in South East Asia.

1. **Chilling injury:** Storage at 5°C for one week, Storage at 10°C for two weeks.
2. **Uneven fruit ripening:** Imbalance in nutrition; Application of Calcium Nitrate 2kg/tree one month before harvest.
3. **Wet core (water core):** caused due to rains just before harvesting.
4. **Flesh burn:** Boron deficiency: application of Borax @10g/tree/year.

Fig (*Ficus carica)*

i) **Sunburn:** Direct sunlight and heavy pruning of the trunk and branches.

ii) **Fruit splitting or cracking:** Sudden charge in climatic condition.

iii) **Fruit drop:** Excessive drought and heat, cold night and light frost results in fruit drop in fig. The spray of GA3 @ 3ml/litre.

Grape (*Vitis vinifera*)

1. **Barrenness:** Loss of floral primordial.
2. **Water berries:** Excessive application of nitrogen fertilizers potash and oil cakes: spraying of boric acid (0.2%).
3. **Shot berries:** Major problem in Beauty Seedless and Perlette.

   Deficiency of boron and Zinc, It can be cured by application of Ethephon (25ppm) + Sevin (2000ppm)
4. **Hex and chicken disorder:** Deficiency of Boron and Zinc
5. **Uneven ripening:** Common problem in Bangalore Blue, Bangalore Purple, Beauty Seedless and Gulabi grapes. Controlled by application of Ethephon 250ppm
6. **Blossom-end-rot:** Defective calcium nutrition Treatment with 1.0% calcium nitrate.
7. **Interveinal chlorosis**: Deficiency of Mg, Zn and Fe, 0.2% sulphate salts.

8. **Pink Berry:** Thompson Seedless and Tas-a-Ganesh in Maharashtra, caused due excessive use of ethylene.
9. **Bud flower and berry drop:** Major contributory factors are moisture stress, C/N ratio and hormonal imbalance with high ABA and low auxin levels. Deficiency of B, Zn and Auxin, NAA (50 ppm)
10. **Berry shrivel:** GA3 – 1.6gm/acre
11. **Cluster apex wilt:** Excessive N supply
12. **Stalk Necrosis:** Deficiency of calcium
13. **Berry cracking and rotting:** Excess of water.

    Dormex – 0.2% after pruning to break dormancy in mid-January.

Guava

1. **Fruit drop:** 45-65% loss, Spray of GA3
2. **Sun Scald:** Direct exposure of fruit to sunlight
3. **Bronzing:** Zn deficiency, L-49 is more tolerant than Allahabad Safeda.
4. **Fatio:** Deficiency of organic matter; N, Zn, B.

Hazelnut

1. Kernel black tips:
2. **Blank nuts:** Defective embryo, unviable eggs and failure of fertilization or embryo abortion.
3. **Brown spots in the Kernal Cavity (BSKC):** found in Spanish hazelnut cultivar, BSKC com effect – 7 to 97%.

Jackfruit

1. **Premature fruit drop:** Lack of pollination and imbalanced nutrition
2. **Chilling injury:** Exposure to temperatures below 12°C (054°F)

Jamun (Fruit of the Gods)

1. Flower and fruit drop: First drop – 52%. Two sprays: GA3 – 60ppm

Kiwifruit

1. **Sun scorch or sunscald:** direct exposure of fruits to sun rays
2. **Water Strain:** due to deposition of tannins

Litchi

1. **Fruit cracking:** Changes in atmospheric Temperature, RH and soil moisture. Spray of Zinc sulphate – 1.5%
2. **Fruit drop:** Zinc sulphate – 1.5%
3. **Sunburn:** Fruit contact with sunlight

Mango

1. Malformation:

   Resistant varieties: Bhadaurem, Alibi, Illaichi. Due to low temperature

   Bombay green – highly susceptible Borax spray – (0.6%)

2. **Blacktip:** Coal fumes and gases of the brick kiln. First reported by the Woodhouse in Bihar 1909. Caused due to gases such as SO2, CO, CO2 and ethylene.

   Bordeaux mixture – 2:2:250 or 1.5 kg of copper oxychloride per 500 litres of water should be used.

3. **Spongy tissue:** First noticed in Alphonso mangoes by Cheema and Dani 1932. Treatment with Heat convection.
4. **Internal necrosis:** B. deficiency, Highly susceptible cv Dashehari.

   Free – Neelam and Langua. Spray of Borax is effective.

5. **Soft nose:** First reported in Florida in an Indian origin variety from Mulgoa due to high nitrogen and calcium.
6. Alternate bearing. Alphonso is highly susceptible to Cultar (PBZ) application 2000ppm.
7. **Jhumka (clustering):** low temperatures in Feb-Mar + improper pollination and fertilization. 200-300 ppm NAA during November.
8. **Leaf scorch:** Chloride toxicity, excess K.
9. **Taper tip:** Deshehari is susceptible.
10. **Tip pulp:** The fruits of Lucknow Safeda are most affected by this disorder.
11. Girdle necrosis:
12. **Fruit pitting:** Dashehari is most effected, B. deficiency. Borax @400-500g/ tree.
13. **Jelly seed:** Nitrogen is reduced.

Mangosteen (*Garcinia mangostana*) Queen of fruit

1. **Gamboge:** Discoloration of the pericarp and the fruit pulp. (Yellow spot)
2. **Flesh translucency / transparent flesh disorder (TFD)**: Excess water during the pre-harvesting stage of fruit development (around 9 weeks after bloom) after fruit cracking.

Olive (*Olea europaea*) Symbol of Peace and goodwill

1. **Fruit drop:** controlled by optimum soil moisture.
2. **Chilling injury:** Storage disorder (caused due to low temperature)

Papaya

1. **Pulp gelification:** Occurs due to magnesium and calcium deficiency.

Passion fruit

1. **Chilling injury**: Keep the fruit above 5°C (41°F)
2. **Heat injury:** Direct sunlight should be avoided by shading the vines.

Peach

1. **Splitting of fruit and greening:** Due to heavy rainfall after a long dry spell.
2. **Internal breakdown:** Storage of control temperature condition.
3. **Inking or black staining:** Affecting only the skin of a peach. Heavy metals and iron, copper and aluminium.

Pear

- **Brown heart:** Caused by abnormally cool growing season preceding harvest temp low or than 7.1°C night and day temp. 21°C.
- **Core breakdown:** high fruit calcium content has been correlated with reduced susceptibility mostly in European pears.
- **Internal browning:** Worldwide problem of Asian pears.
- **Water breakdown of European pear:** Especially the Bartlett variety during storage and ripening. (less than -1°C (30°F)
- **Flesh spot decay (FSD) of Japanese pears:** Low crop load and large fruit.

- **Hard end:** Manifests in the rootstock. Control using European pear as rootstock.
- **Rink calyx or pink end:**

Pecan nut

- **Rosette: Caused** due to expensive crinkling: Zinc deficiency (Z. sulphate 0.5%)
- **Mouse-ear:** Manganese deficiency.

  Persimmon (Introduced by USA)
- **Fruit drop:** Due to lack of pollination early June.
- **Calyx cavity:** Due to excessive application of N and K fertilizers.
- **Skin russeting:** Presence of excessive nitrogen

  Pineapple **(Pineapple is the second most popular tropical fruit next to banana)**
- **Blackheart or endogenous brown spot (EBS):** harvesting chilling temperature generally below 7°C (45°F) for one week or longer.

Plum

1. **Heat spot or Kelsey spot:** This problem has been reported in European plum due to high temp and B deficiency, Borox @500ppm in Sep.
2. **Internal breakdown or chilling injury:** Due to leatheriness or mealiness.

Pomegranate

1. **Fruit cracking:** Sudden change in the moisture content, excessive nitrogen and deficiency. Control by the application of NAA – 100ppm, 100-200ppm GA30.8% Borax.
2. Husk Scald:
3. **Chilling injury:** below 5°C (41°F) during storage.
4. **Internal breakdown:** loses upto 50-60

Raspberry

- **White drupelets:** high temperatures above 40°C

Sapota

- **Wilt:** Due to anaerobic conditions in monsoon and post-monsoon season.
- **Oblong fruits:** Due to high temperature, rainfall and during flowering.

- **Fruit depression:**
- **Corkiness**: Caused due to exposure to intense sunlight due to killing of hydrolyzing enzymes.

Strawberry

1. **Albinism:** Excessive fertilization
2. **Fruit malformation:** Excessive application of nitrogen fertilizer.

Walnut

- **Oil rancidity:** Postharvest physiological disorder.

Other

- Apple helps in curing heart disease, weight loss and cholesterol reduction.
- Wild apricot, popularly known as Zardalu, appears to be indigenous to India. It is an excellent source of Vitamins A, C and E.
- Avocado, is also known as alligator pear or butter fruit. Edible portion flesh, introduced from (Sri Lanka), 20th century.
- The Arabic word "Banana" means finger cashew nut. It was introduced in Goa by the Portuguese in 16th century.
- Traverse City is known as the Cherry capital of the world,.
- Sweet cherry was introduced from Europe before India got independence in 1947.
- Citrus fruits have originated in South East Asia; particularly in India and China.
- The scientific name of Coconut is *Cocos nucifera* where Nucifera means Nut bearing dried coconut.
- Durian is called the 'King of all fruits'. Durian has been derived from the Malay word 'Duri' which means thorns.
- Grape is also known as the "Food of Gods"
- Hazelnut is also known as a cobnut or filbert nut according to the species; it contains 60.5% Fat.
- Jamun is known as the 'Fruit of the Gods'
- Kiwifruit is a natural source of lutein and zeaxanthin (30-35%).
- Litchi is commonly referred to as the "Fruit of romance in China. Great source of Vitamin C, Calcium and Potassium; India is the second-largest litchi producer in the world.

- The olive tree is regarded as the symbol of 'Peace and Goodwill' Olive trees are known for their longevity and can live for 500 years; they can survive drought and wind.
- Admiral Perry who discovered the fruit growing on the coast of Southern Japan in 1851 introduced Japanese persimmons into the USA from Japan.
- The English name 'Raspberry' comes from the old English term 'rasps'.
- Fruit crops contribute to around 29.5% of Agriculture GDP
- Fruit crop cultivation contributes to around 29.91% of the total area occupied by Horticulture
- The term 'Plant physiology' is derived from the Greek words physic meaning nature and logos meaning discourse.
- CAMC (Crassulacean Acid Metabolism) Stomata open early morning/night. Therefore the application of C2H4 during evening /night improves the absorption of hormone in the plant system.

Boron	Deficiency	Zinc	Deficiency
Fruit necrosis	- Aonla	Little leaf	- Any fruit
Fruit cracking	- Bael	Citrus decline	- Citrus
Peel splitting	- Banana	Frenching	- Citrus
Fruit cracking	- Citrus	Start Berry	- Grapes
Crown choking	- Coconut	Bronzing	- Guava
Flash Bume	- Durian	Rosette	- Peanut
Hen and Chicken	- Grapes	Rossetti leave	- Apple
Bud, flower & berry drop	- Grapes	Black end and Dark spot	- Pear
Internal necrosis	- Mango	Multiple crowns	- Pineapple
Fruit pitting	- Mango	Abnormal fruit size	- Cherry
Heat spot & Kelsey Spot	- Plum	Bundy top crowns	- Banana
Water core	- Apple	**Molybdenum Deficiency**	
Blacktip	- Mango	Yellow leaf spot	- Cashewnut
Crown chocking	- Coconut	Yellow leaf citrus	- Citrus
Barren Nut	- Coconut	**Magnesium Deficiency**	
Blossom blast	- Pear	Internal chlorosis	- Grapes
Cork development	- Apple	Pulp gelification	- Papaya
Calyx and rot	- Pear	Internal chlorosis	- Apple
Corky tissue	- Pear	Intern clououis	- Cherry
Calcium Deficiency	High Temperature		
Scald	- Apple	Yellow pulp	- Banana
Cork spot	- Apple	Pit burn	- Apricot
Bitter pit	- Apple	Black scald	- Date palm

Stalk necrosis	- Grape	Darkening	- Date palm
Blossom end rot	- Grape	Skin separate or puffiness	- Date palm
Clax and drop	- Persimmon	White drupelets	- Raspberry
Soft nose	- Mango	Oblong fruits	- Sapota
Manganese deficiency	Low temperature		
Mouse-ear	- Pecan Nut	Blackheart	- Pineapple
Nitrogen Deficiency	Gel Breakdown		- Apricot
Soft Nose	- Banana	Malformation	- Mango
Copper Deficiency	Jhumka		- Mango
Exanthena /	- Citrus	**High humidity**	
Anmorious	Blacknose		- Date palm
Wither tip	- Pear	**High Co2**	
	Brown heart		- Apple

5

Canopy Management in Fruit Crops

Introduction, importance and scope of Canopy management

- The tatura trellis uses more water than other systems.
- Light interception ranges from 57 to 81% of available light.
- HDP increases light interception of 20% and 60%.
- Meadow orchard system affords more light than other conventional systems.
- The orientation of North to South intercept should be about – 89% of direct and 87% of diffused light.
- Summer priming in combination with dormant priming reduces light interceptor by 15%.
- In the bottom of 4m tall trees, the light is reduced to 10% of full sunlight.
- Light transmittance is a logarithmic function of the leaf area index.
- Photosynthetic Active Radiation (PAR) is from 380-720 nm.
- Photosynthesis to be about 60% of the maximum at 25% of full sunlight.
- Tall east-west hedge-rows intercepted more solar radiation than North-South hedgerows at higher latitude during early and late summer.
- The maximum field in dwarf-pyramid trees (180mt/has) in golden delicious (summer).
- Best colour in apple develops with light expcsure of femora then 70% of full sun.
- Adequate colour from 40-70%, inadequate colour with less than 40% full sun.
- Palmate leader is the modification of the central leader system.
- Dwarf pyramid with the lowest branch at 30 to 35 cm from the ground.
- Slender spindle bust, a modification of dwarf pyramid in heading back 60 cm with the lowest shoot at 30 cm above ground level.
- Trellis canopy: developed by IRI from Tatra Australia 1973 of row V-shaped system.

- Multi-rows and bed system: 20000 to 100000 trees/ha.
- Growth retardant: Paclobutrazol, promalin, CCC, MH, AMO 1618, XE1019. BAS-125 (Apage).
- Apple, Training, 1m height of ground level.
- 'Lincolan Canopy'(T-Trellis) was developed in 1970 by Dummon Stop.
- Summer pruning, increases light intensity and red colouration of apple fruits.

  Peach: Palmette – 149/t/ha
- Tatura trellis – 143/t/ha Lincoln Canopy – 119t/ha Central ladder – 109t/ha Vase canopy – 86t/ha.
- In peach, maximum fruit weight was obtained on training angle between 15° and 45° (crotch angle).
- First summer pruning experiment of the peach was conducted in the USA in 1912.
- Sweep Cheng is tall, attaining a height up to 18m, or above and pyramidal found or ground.
- Plum trees are mostly trained on open centre leaders and modified centre canopy.
- Japanese plum is more adaptable to the open centre system.
- Tommy Atkins has ¼ of the centre 8th canopy remove during March.
- Summer pruning grape: March-April in the states of AP, KN, MH June in the states – TN
- Kiwi fruit: Kniffenal Pergola system is commonly followed for training.

NIPER – (National Institute of Pharmaceutical Education & Research established on 26 June 1998.

- Headquarter – Mohali (Punjab)
- Livestock sector form (Methane CH4) (42-44%)
- Cloned Buffalo Calf (Garima) is the second variety by NDRI – 6 June 2009.
- NFDB – National Fisheries Development Board (September 2006).
- The first book in the world on fruit culture of Litchi.
- First in milk production, second in fruit and vegetable products and IIIrd in food grain production at world level.

6

Breeding in Fruit Crops

Breeding objectives genetic resources and techniques for breeding

- There are countries that consume 100 to 150 kg yearly per capita while others which do not even make it to 10kg.
- Genetic variability is the raw material of the plant breeder.
- International network for the improvement of Banana and Plantain (1 NIBAP), Prance to conserve the available germplasm.
- Only pathogen gene-free planting material can be used for construction or distribution from the gene bank for further multiplication or utilization in any breeding program.
- Fruit breeders have therefore tended to utilize single gene variants in performance to polygenic variation.
- In general, for most commercial requirements, homozygous is inadequate by six or seven generations of selling (56 or 37).
- In the dominance hypothesis, it is assumed that the deleterious gene is normally recessive.
- The simplest approach for the selection of our breeders is mass selection.
- Mutation breeding: (Y and X-rays) (l-particles and neutrons), non-ionizing radiation (UV-B light).
- Chemical: Ethyl methanesulfonate, diethyl sulphate, nitroso compound (nitrous acid)

Mango

- India has been growing mangoes for the last 4000 years.
- At present, at least 94 countries produce mangoes.
- Maximum mango yield (33t/ha), pear – 90t/ha, peach-56t/ha.
- Apple – (112t/ha), Orange – 80t/ha, Prunus-45t/ha
- The average yield of mango is8.1t/ha in India.
- *Mangifera indica* L – (Indo-Myanmar region)

- Genes consists of 62 species while (Mukherjee 1972).
- Genes consists of 69 species while (Kostermans and Lampard 1993)
- October is the time of flower bud differentiation in Mango.
- Dec to Feb in the CU. Haden.
    - Aug to be flower bud differentiation in Punjab.
    - Alphonso: ripens early October and reaches the peak by November. Last week of December is critical time for flower differentiation under the northern India climate.
    - *M. caesia* – White pulp, sweet, fragrant
    - *M. decandra* – Rootstock for waterlogged conditions.
    - *M. gedebe* - Rootstock for waterlogged conditions.
    - *M. indica var* Mekongensis – Fruits twice a year and hence a good parent for crossing.
    - *M. apocarpous* – Rootstock for waterlogged conditions.
    - *M. pajang* – Fruit can be peeled like a banana and can yield new varieties by selection.
    - *M. similes* – good for breeding of stone free mangoes.
- In equatorial regions, flowering often occurs throughout the year, this indicates mangoes day natural habit.
- Flower buds in many trees start emerging into rudimentary panicles in the latter half of January and continue to do so till February under India condition.
- Flowers start opening early in the morning and complete in the fore noon.
- Maximum opening is reported between 9-10 cm.
- Stigmata receptivity has been found to continue up to 72 hours after anthesis. Maximum germination of pollen graph on the stigma is recorded at 10 am and 12 noon.
- The stigma remains receptive even up to 5 days after anthesis depending upon the environmental condition.
- The mango pollens remain viable for 10-20 days to room temp.
- Pollen viability is determined by Acetocarmine test at the time of dehiscence is very high (>90%)
    - High pollen viability: Chausa – 94.3% Krishnabhog – 91.4%
    - Langra – 90.8%
    - Dashehari – 93.2-94.0%

- About 10 to 15% pollen germination is obtained in 25% sugar solution to which 0.5% agar is added at a temperature cf 24-27%.
- The mango inflorescence (or panicle) varies widely in length from 15-60 cm and bears mainly two types of flowers: mate and hermaphrodite.

Cultivar	**Perfect flower %**
Langra	77.90%
Dashehari	68.80%
Neelum	16.41%
Romani	0.74%
Chausa	42.90%
Bonbay green	9.2%
Pazli	14.90%
Mallika	25.60%
Tommy atking	39.00%
Kenringtou	7.00%
Mulgoa	6.00%
Lucio-I	4.00%

- Fruit-bearing in bunches has been observed to be dominant over single-fruit bearing.
- Biennial bearing is dominant over regular bearing.
- Fruit size is an important character which is governed by Polygenes.
- Amrapali has a very short juvenile phase.
- Resistance to floral malformation is controlled by recessive genes.
- Spongy tissue, a physiological disorder of fruits, has also been found to be governed by the recessive gene.
- Regular bearing cultivar – Bangalore (Totapuri) and Neenulns.
- Kurukkan can be utilized for imparting dwarfing in the progenies.
- Neelum is highly susceptible to bacterial canker.
- Physiological resistance to canker is found in Bombay green.
- Tommy Atkin is tolerant to – anthracnose.
- Self-incompatibility in Dashehari, langra, Chausa and Bombay green.
- Dashehari is a cross incompatibility between Chausa and Safed Molihnash.
- Bombay green and Dashehari are the best pollinizers for Dashehari and Chausa.

- An ambient temperature of 20% during pollination and fertilization greatly improves seed set.
- Temperatures between 15-20° are generally ideal for controlled hybridization.
- Muslin cloth hybridization work was first initiated in India 1911 at Poona.
- Niranjan has also been identified from Parbhani (MH), bears fruits in October.
- The dwarf cultivar Rumani appears to be a useful donor.
- Clonal selection Paiyar-I Neelum was made at the fruit research station Paiyur, Tamil Nadu.
- A higher phloem/xylem ratio has been shown associated with dwarfing cerotype.
- They treated scions with Y-ray and chemical mutagenic (EMS and NMU).
- Banana germplasm collections in India:

Centre	Ecological condition	Number of Accessions
TNAO	Tropical	243
KAU	Humid Tropical	257
AAU	Humid Tropical	95
GAU	Sub-Humid tropical	65
APAU	Humid Tropical	40
Rajendra AO	Semi-acid subtropical	115
Indian IHR	Mild subtropical	250
NRCB Trichy	Humid Tropical	597
NBPGR Trichy	Humid Tropical	412

- When plants are grown through suckers are about 8-10 months old and produce 25- 50 leaves.
- There are from 10 to 12 leaves still inside the pseudonyms at the flower in the ignition so that the young inflorescence is compact and conical.
- The basal (proximal) notes bear female flowers and those may be from five to eighteen of there nodes.
- The upper (distal) modes contain mate flowers which remain tightly enclosed in bracts forming a conical structure called the "bell".
- These are hermaphrodite flowers which have short ovaries and do not develop into edible fruits.
- Some diploid and triploid parthenocarpic cultivars are fertile. The fertility depends on the presence of viable pollen of pollination.

- In the cultivars of the 'Cavendish' subgroup (AAA), female fertility is absolute.
- Cvs. Dwarf Cavendish (AAA), Robusta (AAA), about 5000 pollen grains are produced of which 4-5% are found viable in the laboratory.
- *M. acuminate* and *M. balbisiana:* 40,000 to 50,000 pollen grains per anther with 60-65% viability found National diploids of bispecific origins (AB).
- Edible diploid cultivars of AA genome, the highest germinability of 45.3%.
- The most effective time for pollination is between 7:00 am 10:30 am.
- They are sown at once for germination or their embryonic are excised aseptically about 2 days later and cultivar in vitro.
- Gross Michael has a characteristic very low level of female fertility while high gate.
- Average seed set ranges from less than 1 to more than 20 seeds per bunch.
- The low seed setting is due to sterility of either pollen or while or both.
- With the drying of the nectar, the ovary apex often becomes necrotic at the time of flower opening and pollen tube failure has been observed in this region.
- Seed germination is pocr and erratic.
- Fresh banana seeds that are high is moisture content germination readily, but after drying they become dormant.
- Properly dried banana seeds remain viable for a few months to 2 years.
- The seed germination ranges from 16.9% in *M. balbisiana* BBX Robusta (AAR) to 96.3% in Matti (AA) M. acuminate (AA).
- The edible banana is mostly from both male and female lexes.
- Parthenocarpy does not appear to occur in *M. balbisiana* there being mo edible BB or BBB types.
- Seeded non-edible Banana 'Calcutta-4' lacks the P1 gene and is homozygous for P2 and P3.
- The inheritance of apical dominate in plantain is determined by gene (ad.).
- The dominant allies Ad, which is probably fixed in AA banana improves the suckering.
- The rate of sucker growth is the result of Gibberellin (GA3) level regulated by Ad genes.
- Dwarfism in planting structure is controlled by a single dominant gene in (AAA).

- 'Bobby Tannap' is due to a recessive gene 'dw'.
- Balbisiana (B) genome confers headliners to drought, disease resistance and acid-starchy fruit quality.
- *M. acuminate*, spp. maucconsis or *M. acuminate* spp. burmlanvi, sigatoka leaf sport pathogen or Panana wilt fungus is controlled by several dominant genes.
- Sigatoka is under the control of one excessive allele (bs) and modifier gene action of other few independent resistant (additive) alleles bsr2 and bsr3.
- Panama wilt was under the control of a dominant gone is tetraploid offspring obtained by crossing the susceptible 'Gros Michal' with diploid accessions.
- SH-699 is resistant to bacterial wilt.
- Host-plant response to weevil (cosmopolites sordidus) in Musa is controlled by gene(s).
- Banana breeding started as early as 1922 in Trinidad and 1924 Jamaica.
- In India, Breeding work started at the CBRS, Aduthures (T.N) in 1949.
- Mutation: Highgate – (AAA) locos (AAA), Gras Michael (AAA) Motta Poona (AAB) is a sport of Poonam (AAB).
    - Agiranka Rasthali is a sport of Rasthali. Barhari
    - Malbhog is a spot malbhog.
    - Krishna Vazhai is a – Virupakshi
    - Sombromi Monthan (ABB) – Monthan (ABB)
- Soaking seeds in 0.11 MEMS solution for 48 hrs at 20°C reduces germination by 50% but has been practiced for inducing nutrition in banana.
- The optimum dose of Y-radiation from a 6°C sources varices from 25 to 60 kg depending upon ploidy level of the genotype.
    - 25 kgy – diploids
    - 35 kgy – AAA triploid
    - 40 kgy – AAB and ABB triploids 50 kgy – AAAA tetraploids
- An early flowering mutant of 'Grande Naine' has been produced by irradiation at international atomic energy agency in Australia and given the identification number GN-60A.

Citrus

- The trifoliate leaf character of Poncirus shows essentially complete dominance over the metabolite condition of citrus.

- Resistant to citrus Tristeza cluster virus (CTV), is controlled by dominant alleles at a single locus (ctr)
- Resistant genotypes were designated as Ctr-Rr and Ctr-RR and susceptible scion as Ctr-rr.
- The Hong Kong wild Kunnquat (*Fortunella hindsii*) (Champ)
- The international production of tetraploids by the use of colchicines (C22H25NO6) or other treatments has received little attention.

Papaya

- M. dominant for maleness M2 dominant for hermaphroditism and M recessive for femaleness.
- The combinations M1M1, M2M2 and M2M1 are lethal and thus fail to produce viable seeds.
- M1m gives male trees, M2m hermaphrodite trees and mm female trees.
- Mp stands for male plants.
- Mp for the hermaphrodite tree.
- A dominant gene for suppressing femaleness (50F) is present only in the pure male.

7

Hi-Tech Horticulture

Introduction to Hi-Tech Horticulture

Greenhouse area in different countries

1. Japan – 42,000 (ha)
2. Holland – 10,000 (ha) 3) Italy – 20,000 (ha)
4. USA – 4,000 (ha)
5. India – 700 (ha)
    - Crops are grown in the greenhouse
      Strawberry, grapes, citrus, banana, papaya, melons.
    - Salt affected soils in different states: UP - 2,375
        - Gujarat – 1,334
        - W.B. – 1,150
        - RJ – 458
        - Punjab – 749 Haryana – 636

	pH	Ece (dem.)	ESP	SAR
Saline Soil	<8.5	>4	<15	<13
Saline Sodic	<8.5	-	-	-
Soil	+04 variable	>4	>15	>13
Sodic soil	>8.5	<4	>15	>15

- Saline sodic soils: Chlorides and sulphate of Na, Ca and Mg excluding gypsum.

Soil sterilization Physical agents

A. Heat (high temp)

a) Dry heat

i) Plame sterilization (direct heat)

ii) Hot dry coir (indirect heat) hot air oven

b) Moist heat (steam)

   i) Stream under pressure (Autoclaving): Autoclave

   ii) Stream without pressure (Arnold steam sterilize)

B. Radiation

   i) Ultraviolet light

   ii) X-rays

   iii) Gamma rays

   iv) Cathode rays

C. Filtrations

   i) Sintered glass filters

   ii) Asbestos pads or Seitz filters

   iii) Membrane filters (Millipore filters)

   iv) Air filters (laminar flow combinate)

Chemical agents

i) Phenol and phenolic-phenol-5% (carbolic acid).

ii) Alcohol-ethyl alcohol, ethanol 7 to 9%.

iii) Halogens – Lodine (tine tune) 2%

iv) Chlorine and chlorine compounds: Sodium hypochlorite 1%

v) Heavy methods – Mercuric chloride: 0.1%

   o Silver nitrate: 0.1%

   o Copper sulphate: 5%

vi) Aldelydes – Formaldehyde (Farmalin)

Gaseous agents

i) Ethylene oxide

ii) Propylene oxide

iii) Formaldehyde

iv) Methyl bromide

v) Ozone

   - Greenhouse: X The life of the UV stabilized (200 microns) film is about 2-3 years. (Polyethylene)

- (Polyvinyl): Cost of the films is 2-3 times higher than polyethene (up to 5 years)
- Polyester (lost up to 15 years) The cost is about three to four times higher than that of polythene.
- The life expectancy of several cladding materials:
- Life expectancy (years)
  1. Glass – 30+
  2. Double polythene / UV treated – 3-5
  3. Acrylic SDP – 15+
  4. Polycarbonate SDP – 20+
  5. Fibre glass – 15-20
- The cost of shade net is Rs. 18-21/m2 as per the colour and shade %. Parthenogenesis is a form of apomixes.

Protected cultivation

- The production of crops in greenhouses is sometimes called plant forces.
- The optimum level of R.H in a greenhouse 80-90%.
- The greenhouse that runs from east to west is the best.
- Fibre glass sheets of UV resistant polyethene film may be used as a covering material.
- Polyethene sheet of 0.10 to 0.15mm thickness is resistant to UV ranges.
- Ideal size of greenhouse 100 to 500sq.m.
- Hotbeds can be used throughout the year except in areas with severe winter.
- The rooting of softwood leafy cuttings under spray or mist is a technique now widely used by nurseries.
- Mist propagation units are used for propagation of "difficult to root" cutting in most advanced countries in the time usually 6 seconds on and 990 seconds off.
- The optimum pH of the water to be used in the mist unit is 5.5 to 6.5.
- The development of blue and green algal growth is very common in mist propagation.
- Net houses are widely used as propagation structures in tropical areas, where artificial heating is not required and artificial cooling is expensive.

- Bottom heat box: It was developed at IARI. The ideal temperature to be maintained in the box is 30+2°C be at the temperature for cuttings of mango and grape root easily and profusely. In general, it takes 1-2 months for proper development of the root.
- Plastic tunnels: It has been developed for greenhouse crops research in little Hampton, England. The same is now being used by nurserymen through the world.
- Propagation media should be highly decomposed (20C: 1N) to prevent N immobilization and formation of an excessive sheet during production. It should have a high cation exchange capacity (CEC), Ideal pH (5.5-6.5). In soil having 40% sand, 4% silt 20 days. Sand 0 05-2mm is diamond.
- Moist peat: It has a high moisture holding capacity (15 times its dry weight and a high acidity (pH of 3.2-4.5). This type of peat generally comes from Canada, Ireland or Germany.
- Peat moss is the most common use of peat in horticulture in the course of being the best.
- Compost gets ready in 12-16 months.

8

Mango

Common Name	-	King of fruit, Bathroom fruit
Botanical Name	-	*Mangifera indica*
Family	-	Anacardiaceae
Edible Portion	-	Mesocarp
Chromosome No (2n)	-	40, X - 10
Climatic adaptability	-	Tropical
Fruit morphology	-	Drupe
Rate of respiration	-	Climacteric
Relative salt tolerance	-	Highly sensitive
Self-incompatibility	-	Homomorphic (Sporophylic)
Type of pollination	-	Cross-pollination (Housefly)
Origin	-	Indo-Burma
Ploidy level	-	Allo-tetraploid / Amphidiplid
National Fruit of India	-	Mango
Area	-	2258 Thousand ha.
Production	-	21822 Thousand MT.
Productivity	-	9.66MT/ha
State with maximum area under cultivation	-	Uttar Pradesh (265.62 Thousand ha.)
State with maximum production	-	Uttar Pradesh (4551.83 Thousand MT.)
State with maximum Productivity	-	Rajasthan (17.58 MT/ha)
Harvesting period	-	February - Mid -August
Storage temperature	-	8-9°C
Storage life	-	Cold storage (35 - 45 days)
Yield (MT/ha)	-	8.0, Hybrids – 15-25

Propagation method	-	Veneer grafting
Spacing	-	10 m × 10 m
Planting time	-	June -July
Largest producer in the world	-	India (54.2%)
Natural habitat	-	Alternate bearing
Established	-	Central Mango Research Station (CMRS) Lucknow- 1972
Mango	-	Polyembryonic
Inflorescence	-	Panicle
Ethylene production	-	Medium (1-10) (µl C2H4/Kg/hr)
Respiration rate	-	Medium – (10-20mg) (Release of CO2 Flower)
Bearing habit	-	Terminal (Old season group)
Fruit bud	-	Simple bud
Sex ratios	-	B& high : @& less (1:36%)

Flower bud differentiation period Oct-Dec (But in Dashehari May-June and Sept-Oct)

Percent of flowers developing into fruit	-	0.1%

Largest area under Mango cultivation in India: (39.2%)

Low temperature	-	Highly susceptibility
Two crop of mango	-	Kanyakumari district of Tamil Nadu
First hybridization work start	-	Burma and Prayag (1911 at Pune)
Flowering time	-	North India – February

South India – September

Harvesting period

Kerala	-	February
TN, AP, Maharashtra	-	May
Bengal	-	June
North India	-	July to August
Acidic index (pH)	-	5.5-7.5

Annual rainfall requirement - 75-190 cm

Common names of Malda - Fazli in Calcutta Bombay green in UP Sehroli in Delhi

- The 'Portuguese Jesuits in Goa started Inarching about 300 years ago. (90% success)
- The seedlings can be inarched when they are three months old as is done in the "Philippines".
- The Indian nurserymen usually practice inarching in one-year-old seedlings.
- Grafts onto older seedlings result in poor root system.
- There should be no rain; otherwise, water gets inside the union and causes rotting of tissues.
- The name *Mangifera* was given for the first time by "Bontus" in 1658 when he referred to this as arbor Mangifera (the tree producing mango). Linnaeus also referred it as *Mangifera arbor* in 1747 before changing the name to its present form "*Mangifera indica*" in 1753.

  Mango was firstly grown within glasshouse in 1960 at England.

Region wise cultivars

Northern region	-	Dashehari, Langra, Chausa, Bombay Green and Amrapali
Eastern region	-	Himsagar, Langra, Fazli, Zardalu, Krishnabhog Gulabkhos and Amrapali
Western region	-	Alphonso, Pairi, Kesar Rajapuri, Molkurad, Jamad or Jahangir, Imam Pasand etc.
Southern region	-	Bangaloro, Neelum, Mulgo, Swarnrekha, Pairi (Peter), Banganpalli, Badami (Alphonso)
Bombay green	-	It is an early variety but susceptible to malformation.

- Mango Malformation was first observed in 1891 in Bihar.
- Malbhog variety of mango is most susceptible to the waterlogged condition.
- Dasheheri variety has high fruit retention.
- Mango can withstand deficiency of P but not K.
- Mango variety Mulgoa is mono-embryonic in India and polyembryonic in Florida.
- Highest mango productivity in the World: Brazil

  North Indian variety - Alternate Bearer; monoembryonic, self incompatible

South Indian varieties	-	Regular bearer, polyembryonic
Pollinating Variety	-	Bombay green (highest Vit. C).
Maturity indices	-	Alphonso-1.01-1.02 Specific Gravity (SG) Dashaheri -1.00 (SG)

- Mangoes are highly susceptible to low-temperature injury. Hence, they should be kept at about 5°C during storage as lower temperatures will damage them.
- "VHT" (Vapour Heat Treatment) is recommended for disinfection of mango plant against fruit flies and stone weevils.
- First time caging technique of mango breeding was used by RN Singh.
- Good mango varieties have a TSS of 20°B.
- *Mangifera indica* is closely related to *M. longipes* Giff and *M. sylvatica* Rexb.
- The first systematic record of its varieties was made in Ain-i-Akbari in 1590AD.
- Fotapari was resistant to powdery mildew.
- Dashehri – Nectar making
- Mallika – Canning of slices.
- Caroldrisis was the first to mention mango.
- The polyembryonic seedling of *M. indica* CV. Vellaikumban was found to be 2n=80.
- Dashehari with Neelum yields one fruit per panicle and has bench bearing as a dominant character.
- The incidence of floral malformation in the early flesh of panicles is much higher than the late emerged panicles.
- The amount of RNA, DNA and protein is always higher in the healthy panicle than in the malformed ones.
- **Polyembryonic rootstocks of mango:** Bappakai, Chandrakaran, Goa, Olour, Kurekkan
- **Salt resistant rootstock:** Solan, Mulgoa, Bellary, Vellaikolamban, Nileshwar
- **Dwarf vigorous rootstock:** Moovandan and Nekkere (Salt resistant)
- **Softwood grafting:** This method of grafting was first developed in India on mango at Gujarat Agriculture University, Anand in 1971-72.

- Rumani: Apple-shaped and vigorous rootstock, but dwarfing effect in Dashehari.
- Olour is used for dwarfing effect in Langra and Himsagar.
- Vellaikolamban rootstock for dwarfing and allopolyploid effect in Alphonso.
- Vellaikolamban is an octoploid (2n=8X=80) variety, potential polyploidy dwarfing rootstock of mango.
- Salt-resistant rootstock of mango: Kurukan, Moovandan, Nekhare
- Mango plant produces perfect Bisexual, polygamous flowers/
- Number of perfect Bisexual / polygamous flowers:

  Highest – 68.9% in 'Langra' variety

  Lowest – 0.74% 'Rumani' variety
- Spongy tissue was first observed by Cheema and Dhani in 1934.
- Temperature between 24-27°C is ideal for mango cultivation.
- The introduced polyembryonic rootstock of mango: Apricot, Simmonds, Higgins, Pico, Sabre Strawberry, Combodina, Terpentine, Carabao and Saigon.
- Ambica: Amrapali × Janardan Parsad (Regular bearer yellow colour with red blush suitable for domestic and export market.
- Xavier – This variety has the highest TSS-24.8 Brix.
- **Al Fazli:** It is a hybrid of Alphonso × Fazli released by Fruit Research Station, Sabour. It is superior to Fazli, does not have spongy tissue. Fruits are sweet to taste.
- **Alphonso:** It is one of the choicest varieties in India. It is mainly grown in the Ratnagiri area of Maharashtra and to a small extent in parts of South Gujarat and Karnataka. Fruits are medium sized (250g), have attractive blush towards the basal end. Pulp is firm, fibreless with excellent orange colour. It has good sugar/acid blend. Keeping quality of the fruit is good. It is susceptible to spongy tissue.
- **Ambika:** It is a cross between Amrapali × Janaradhan Pasand developed by Central Institute for Subtropical Horticulture, Lucknow. The fruits of this variety are medium in size with slight sinus and beak and broadly pointed apex. Peel is smooth and tough. Fruits are bright yellow in colour with dark red blush. Pulp is firm with scanty fibre. TSS of this variety is 21° Brix. It is a late maturing variety.
- **Amrapali**: It is hybrid developed by crossing Dashehari × Neelum. Plants are dwarf and have regular bearing habit. Fruits weigh on an average about

180-250 g and are borne in clusters. Fruits are sweet to taste and have good keeping quality.

- **Arunika:** It is a hybrid between the cross Amrapali × Vanraj developed by Central Institute for Subtropical Horticulture, Lucknow. Fruits of this variety are attractive and with red-blush. Fruit have high TSS (24^0 Brix) and high carotenoids content. Pulp is firm. Variety is regular bearer and plants are dwarf in stature.
- **Arunima:** It is from the cross between Amrapali × Sensation, released by Indian Agricultural Research Institute, New Delhi. The fruits are medium sized, having attractive skin colour. The pulp is deep yellow in colour and TSS is around 20°Brix.
- **Au-Rumani :** It is a hybrid developed from the combination Rumani × Mulgoa released by Fruit Research Station, Kodur. It bears large fruits, good flavour, heavy yielder, flesh moderately firm.
- **Banganapalli:** It is a widely cultivated early season variety of South India. It is the main commercial variety of Andhra Pradesh. The fruits are large sized, weighing on an average of about 350 to 400g. The pulp is fibreless, firm with sweet taste and is yellow in colour. Fruits have good keeping quality.
- **Bombay Green:** It is one of the earliest varieties of North India. It is a biennial bearer. The fruits are medium sized weighing around 250g. Fruits have strong and pleasant flavour. Pulp is soft and sweet.
- **Chausa:** It is a late maturing variety of North India, which matures during July or beginning of August. Fruits are large weighing about 350 to 400 g. Fruits are bright yellow in colour. Pulp soft and sweet. It is a biennial bearer.
- **Dashehari:** One of the choicest varieties of North India. It is a mid-season variety with biennial bearing tendency. Fruits are medium sized, with pleasant flavour, tastes sweet, pulp is firm, and fibreless. Stone is thin and keeping quality is good.
- **Fazli:** It is indigenous to Bihar and West Bengal and is a late variety (maturing in August). Fruits are large with firm to soft flesh. Flavour is pleasant and pulp is sweet and fibreless. Keeping quality is good.
- **Fernandin:** It is a variety indigenous to Goa. Fruits are medium sized and are sweet to taste with deep yellow and firm pulp. Pulp is free from fiber. Fruits on ripening get red blush in yellow background.
- **Goa Mankurd:** It is a mid-season variety. Fruits are medium sized, skin yellow in colour. Flesh is firm, and cadmium yellow in colour and fibreless. Keeping quality is good.

- **Gulabkhas:** It is a variety indigenous to Bihar. It is a regular and heavy bearer. It is mid-season variety. Fruits are small to medium sized. It has good rosy flavour and tastes sweet. Fruits are amber-yellow in colour with reddish blush towards the base and sides. Keeping quality is good.
- **Himsagar:** It is one of the choicest varieties of West Bengal. It is a regular bearer. It matures early. Fruits are medium sized with excellent flavour. Taste is sweet. Flesh is firm and fibreless. Keeping quality is good.
- **Jamedar:** It is a famous variety of Gujarat. It is a heavy and regular bearing variety. Fruits are of medium sized having good quality. Flesh is firm, yellow in colour, fibreless with pleasant flavour. Keeping quality is good.
- **Jawahar:** It is a hybrid between the cross Gulabkhas × Mahmood Bahar developed by Bihar Agriculture University, Sabour (Bihar). It is a mid- season variety. Fruits mature from second week of June, precocity and regularity in bearing. The fruits become greenish yellow on ripening. Its fruits are of medium size and average fruit weight is 215g per fruit. Its pulp is light yellow in colour, sweet in taste and pleasant flavoured and remain firm even after ripening. The TSS, acidity and pulp percentage are 22.5, 0.14 and 79.5 per cent respectively.
- **Kesar:** This is a famous variety of Saurashtra region of Gujarat. It is an irregular bearing variety. Fruits are medium sized. Flesh is sweet and fibreless. It has excellent sugar-acid blend. Fruits ripen to attractive apricot-yellow colour with a red blush. It has good processing quality.
- **Kishanbhog:** It is indigenous to West Bengal. It is a mid-season variety. Fruits are medium to large in size. Fruit quality is good and flavour is pleasant and mildly turpentine. Flesh is firm with few fibers. Keeping quality is good.
- **KMH 1**: It is a hybrid released from Fruit Research Station, Kodur. It is from the parentage Cherukurasam × Khader. Plants are semi dwarf, regular bearing, fruits-fibreless has high brix and low acidity.
- **Konkan Raja**: It is a hybrid from the cross between Banglora × Himayuddin. Fruits are small to medium sized, less acidic and best for salad purpose.
- **Konkan Ruchi**: It is a hybrid from the cross Neelum × Alphonso. It is a regular bearer. Fruits are large with thick skin, acidic and preferred for pickling.
- **Langra:** It is one of the important commercial varieties of North India. It is a biennial bearer. It is a mid-season variety. Fruit quality is good. Flesh is firm with lemon yellow colour, scarcely fibrous. It has a characteristic turpentine flavour. Keeping quality is medium.

- **Mallika:** This hybrid between Neelum × Dashehari was released from IARI, New Delhi. It has a strong tendency for regular bearing. The fruits on an average weigh about 350-400g with deep yellow pulp, high TSS, good flavor, uniform fruits and moderate keeping quality (Singh *et al.,* 1972).
- **Manjeera**: This was released from Fruit Research Station, Sangareddy. It is from the parentage comprising Rumani × Neelum. It produces round shaped fruits. It bears regularly and is a prolific bearer, pulp is firm and fibreless.
- **Menaka:** Menaka is selection from Gulabkhas seedling. It is a regular bearing variety; fruits are attractive with deep red basal portion. Pulp is deep yellow, sweet and pleasant in flavour, less fibrous and firm. Fruit shape is oblong-oblique. Average fruit weight is 300g, late maturing variety and the TSS of the fruit is 20 percent, acidity 0.14 per cent and pulp content is 75 per cent.
- **Mulgoa:** It is a late variety of South India and a poor yielder. Fruits are large and round with good quality. Flesh is firm, mustard yellow and fibreless. It has good flavour and tastes sweet.
- **Neelum:** It is a heavy yielding late season variety of South India with regular bearing habit. Fruits are medium sized with good quality and flavour. Flesh is soft, yellow and fibreless. Keeping quality is good.
- **Pairi:** It is native to coastal Maharashtra including Goa and is a popular variety in Karnataka It matures early and is a heavy as well as regular bearer. Fruits are medium sized with good quality. It has good flavour with good sugar acid blend. Flesh is soft, primuline yellow and fibreless. Keeping quality is poor.
- **Pant Chandra:** This is a clonal selection of Dashehari from Govind Ballabh Pant University of Agriculture & Technology, Pantnagar. Plants are tall with erect growth habit. Fruit colour at maturity remains green. It is a mid-season variety. Fruit weighs up to 150g. Fruit pulp is reddish yellow with total soluble solids of 18% and as a pleasant aroma.
- **Pant Sinduri:** This is a clonal selection of Dashehari from Govind Ballabh Pant University of Agriculture & Technology, Pantnagar. Trees are medium in height with round top canopy. Fruit colour is yellow with pink shoulder. Average fruit weight is up to 200g. Fruit pulp is yellow with pleasant aroma. Total soluble solids vary from 16-18% with average yield up to 150 kg per tree. Fruit matures during last week of May to first week of June.
- **PKM-1**: It is a hybrid released from Horticultural Research Station, Periyakulam. the parentage comprises Chinna Suvarnarekha × Neelum. It is regular bearing, produces good quality fruits and bears in clusters.

- **PKM-2**: It is a hybrid released from Horticultural Research Station, Periyakulam. Its parentage comprises Neelum × Alphonso. It is regular bearing and produces good quality fruits in clusters.
- **Pusa Lalima:** It is a hybrid between the cross Dashehari × Sensation developed by Indian Agricultural Research Institute, New Delhi. The plants are semi- vigorous, regular bearer, attractive oblong fruits with bright red peel on yellowish green background with orange pulp and good sugar: acid blend.
- **Pusa Pitamber:** It is a hybrid between the cross Amrapali × Lal Sundari developed by Indian Agricultural Research Institute, New Delhi. The plants are semi- vigorous, regular bearing with attractive oblong fruits. Fruits turn uniform yellow on ripening. It has good sugar: acid blend and uniform size fruits.
- **Pusa Prathibha:** It is a hybrid between the cross Dashehari × Amrapali developed by Indian Agricultural Research Institute, New Delhi. It is regular bearing variety with attractive fruit shape, bright red peel and orange pulp. It has oblong shaped, uniform sized fruits. The plants are semi-vigorous. It has good sugar and acid blend and uniform fruits.
- **Pusa Shresht:** It is a hybrid of Amrapali × Sensation developed by Indian Agricultural Research Institute, New Delhi. Trees are semi vigorous, regular bearing with elongated fruits and attractive red peel. Pulp is orange in colour, fibreless and firm at ripening, moderate sugar: acid blend with uniform fruit size (228g). It contains good amount Beta -carotene and ascorbic acid.
- **Pusa Surya**: It is a selection from the variety Eldon released by Indian Agricultural Research Institute, New Delhi. It bears medium sized fruits, has red peel colour similar to Sensation.
- **Rajapuri:** It is one of the commercial cultivars of Gujarat. It is heavy and regular bearer. Fruits are large sized. It matures during early to mid-season. Flesh is firm, pinnard yellow and fibreless. Keeping quality is medium.
- **Ratna**: It is a hybrid formed by crossing Alphonso × Neelum. It was released by Fruit Research Station, Vengurla. It is regular bearing, produces medium sized fruits weighing on an average about 250 g. Pulp is orange in colour and free from spongy tissue and fibre.
- **Rumani:** This variety produces round shaped fruits having uniform yellow skin colour with red blush. It is grown extensively in the state of Tamil Nadu. Fruits are sweet to taste. Pulp is yellow in colour and becomes soft on ripening

- **Sindhu:** It is a hybrid progeny derived by back crossing Ratna × Alphonso released by Fruit Research Station, Vengurla. Fruits are borne in clusters. Fruits weigh on an average about 150-220g. Pulp is deep yellow in colour and it has good sugar acid blend. Fruits are almost seedless with very thin stone, though fruits above 200 g have well developed seeds.
- **Subhash:** It is a selection from the seedlings of Zardalu. It is a mid-season variety. The ripened fruits are bright yellow in colour. Being similar to Langra fruit in shape, fruits are medium in size with an average weight of 220g. The TSS and acidity of the fruits are 24 and 0.29 per cent respectively. The pulp content is 76 per cent.
- **Sundar Langra**: developed from the cross between Sardar Pasand × Langra. It is regular bearing and resembles Langra in shape and size. Fruits are medium sized and sweet to taste (Hoda and Ramkumar, 1993).
- **Suvarnarekha:** It is a variety indigenous to Andhra Pradesh. It is a heavy and regular bearing variety. It matures early. Fruit is medium, ripens to light cadmium colour with red blush. Flesh is soft and primuline yellow coloured and fibreless. Fruit quality is medium to good. Keeping quality is good.
- **Totapuri:** This variety is grown widely in South India. It is regular and heavy bearer. Fruits are medium to large with prominent sinus. Fruit quality is medium. It has typical flavour and tastes flat. Flesh is cadmium yellow and fibreless.
- **Vanraj:** It is a variety found and grown in the state of Gujarat. Fruits are medium to big sized weighing on an average about 300g. Pulp becomes soft on ripening. Pulp is deep yellow in colour and sweet to taste. Fruits on ripening get attractive red colour.
- **Zardalu:** It is indigenous to Bihar and matures towards the end of June. It is a biennial bearer with medium to heavy bearing. Flesh is firm to soft, capuchin yellow and sparingly fibrous. Fruit quality is good with a pleasant flavour. Keeping quality is medium.

Storage temperature

A. Mature fruit at 6-7°C

B. Ripened fruit at 20°C

- The longevity of mango seeds: 30 days (4 weeks) after extraction or harvest
- The blacktip was first observed in 1909 by Woodhouse.
- In vigorous varieties, application of drenching with paclobutrazol @ 5g/tree/ year is effective in reducing the tree size.

- Mango malformation – The disorder is wide spread in the mango orchards of northern and western India. It has not been observed in Southern India. All other mango varieties except Bhadauran, Alib and Illaichi are resistant to this disorder.
- Control –Spray of NAA 200ppm at the time fruit bud differentiation (October-November in Northern India).
- *Mangifera* contains about 41 species. The maximum number of species reported so far in genes *Mangifera* are 69 by Bumpard and Kosterman 1993.
- Paclobutrazol / Cultar / PP333 / @5g / tree is commercially used for regular flowering and fruiting in Alphonso and Kesar mango in Maharashtra and Gujarat, respectively.
- Alphonso, Kesar, Gulabkhas, Safeda Pasand and Lakhanbhog are mango cultivars suitable for export.
- In mango, there is a heavy drop of hermaphrodite flowers up to 99%. Only 0.1% or fewer hermaphrodite flowers develop the fruit to maturity.
- Spray of 2,4-D @10 ppm reduces fruit drop in mango.
- Parthenocarpy is not common in mango.
- Sindhu is a seedless variety of mango. It results from stenospermocarpic parthenocarpy.
- Dashehari, Langra, Chausa and Bombay Green are self-incompatible mango cultivars.
- Dashehari is cross incompatible with Chausa, Chausa is cross-compatible with Langra and Safeda Malihabad.
- Amrapali and Arka Arena are dwarf cultivars.
- Bombay green is the best pollinizer for Dashehari, Chausa, Bombay green and Chausa.
- Inflorescence in mango is a panicle, male and hermaphrodite (andromonoecious).
- Neelum is a high yielding regular bearing hybrids
- Amrapali (Dashehari × Neelum) was developed in 1978 at IARI, New Delhi, dwarf suitable for HDP (2.5 × 2.5 m2) 1600 plants/ha triangular method. Fruit weight 112.3-143.0g, oblique in shape, â-carctene content (Dr PK Mujamdar)

- Mallika: (Neelum × Dashehari) was developed in 1971 (IARI) Regular bearer, highest Vit-A content, weight 307-600g, resistant to fruit drop, 'elliptical in shape' and β-carotene content.
- Ratna (Neelum × Alphonso) was developed in 1981, Regular bearer free from spongy tissue and fibre, pulp 78.62%, TSS-23.0 Brix, acidity (0.25%) developed by FRS, Vengurla, Maharastra.
- Arunica – Amarpali × Vanraj was developed in 2009
- Sindhu (Ratna × Alphonso) seedless, stone account 3% of total fruit weight, weight 6.75g, pulp 83%, pulp to stone ratio 26:1, Result from stenospermocarpic parthenocarpy, developed in 1992.
- Arka Puneet: (Alphonso × Bangemplli) regular and prolific bearing, free from spongy tissue.
- Arka Aroma: (Banganpalli × Alphonso) dwarf, precious, medium and regular bearer, fruit large, fibreless, good flavour free from spongy tissue.
- Arka Anmol:(Alphonso × Janardan Parsand) semi vigorous, regular bearer, fruits medium, fibreless, good keeping quality and free from spongy tissue.
- Arka Neelkiran: (Alphonso × Neelum) semi vigorous, late bearer, fruit medium, free from spongy tissue.
- Manjeere: (Rumgi × Neetum) precious, regular and prolific bearer, fruit medium, fibreless, sweet.
- Gaurav: (Dashehari × Totapari Hyderabad) was developed from Saharanpur, UP
- Rajiv: (Dashehari × Romani) was developed from Saharanpur, UP
- Saurabh: (Dashehari × Fazl; Zafrani) was developed from Saharanpur, UP
- Rosica – Mutant variety
- Off-season bearer – Niranjan, Madhulica
- Regular bearer variety – Neelum, Gelsbkhas, Himsagar, Pairi, Totapuri.
- Exotic coloured cultivars – Tommy Atkins, Zilette, Haden, Sensation, Julie.
- Lal Sindheeri – Powdery mildew resistant variety.
- Madhulica– The most precious cultivar of mango.
- Neelum – The best combiner variety, ideal for distant transport, two crops in a year.
- Pusa Surya- It is a selection from the variety Eldon released by Indian Agricultural Research Institute, New Delhi. It bears medium sized fruits, has red peel colour similar to Sensation.

- Pusa Aruniina – (Amarpali × Sensation)
- Akshay – selection from Dashehari
- Saisugartha: (Totapari × Kesar) regular bear, free from malformation, suitable for pulp making.
- Coloured mango cultivars: Tommy Atkins, Zilette, Haden, Sensation and Julie are introduced from Miami, Florida (USA).
- Floral Biology – Flower bud differentiation has been reported to start in October.
- Parthenocarpy in mango can be induced by the application of Benzyl adenine (BA) @200 ppm at anthesis and GA3 (250 ppm) + BNOA (10 ppm) at a fortnight internal thereafter.
- Rootstocks of standard size (22-22.5 cm) above ground for grafting
- Mango has high viable pollen collected between 8 to 10 in the morning.
- Best time for Pollination– 10-12 noon.

Months	Flowering
Nov-Dec	Andhra Pradesh
Feb-Mar	Northern India
Jan-Feb	Eastern part India
June-Oct	Queensland and Australia
July	Fiji
Nov-Jan	Egypt

- Duration of flowering – 2-3 weeks
- In mango, more than 50% of the flowers do not receive any pollen grain.
- Fruit development of Langra and Deshehari starts in the last week of March and is completed by the second week of June. According to the percentage, least growth occurs in May while maximum increase in weight and volume occurs in June.
- The carbon/nitrogen ratio in selection to flowering in mango cv. Langra was first studied by Naik and Shaw (1937).
- Spraying with 1% to 2% KNO3 and 6% to 18% Calcium Nitrate helps to promote flowering and fruit set in mango.
- Applying NAA to thin flowers helps promoting moderate cropping in the off-year.
- Pruning of fruited shoots is useful in promoting lateral flowering in the off-year.

Irrigation: The total annual water requirement of the nature mango tree is approximately 11, 976 m3/ha while the seasonal water requirement varies from 20m3/ha/day in winter, to about 443m3/ha/day in summer. However, one must stop irrigation at least 2-3 months before flowering period for obtaining good flowering.

- One-year-old plants 75gN, 110gP2O5 and 55gK2O and 40-80 per cent of nitrogen should be in the form of organic manuring.
- The spray of the combined application of $ZnSo_4$ (0.4%) and urea (1%) produces highest fruit number and yield. Fruit quality is improved by the application of H_2BO_4 0.4% + urea 1% to young mango plants.
- **Grading** – Mango fruits are grouped under three categories on the basis of the fruit weight namely:
- Grade Weight

Grade		Weight
Grade I	:	320g weight
Grade II	:	170-320g
Grade III	:	170g

- **Ripening:** The fruits are spread over paddy or wheat straw in single, double or triple layers at a temperature range of 19.4° to 21.1°C in a ventilated store.
- Dashehari mango treated with GA3 @ 200mg/ litre hastens ripening by 3 days.
- Amrapali fruits require treatment with GA3 but with 50-75ppm and ethereal 500 ppm as pre-harvest spray.
- Ten-year-old plant requires 3kg urea + 3kg SSP + 1kg MOP + 50 kg FYM. The dose will remain constant thereafter. Such plants produce 200-250 fruits per tree.
- Fruit maturity period of different cultivars varies from 83 to 121 day during the 'On' season and from 109-135 day in the 'Off' season.
- These baskets contain about 20kg of fruit. In the western and southern India, bamboo baskets are used in which up to 100 fruit are packed between layers of straw or grass.

Global Producers: India>China> Thailand > Mexico>Indonesia

State wise Producers: Uttar Pradesh>Andhra Pradesh>Bihar>Tamil Nadu> Karnataka.

Physiological disorders

Black Tip: Reported by Woodhouse in 1909, it is caused by the gases CO2, SO2 and acetylene emitted from the brick kilns smoke.

- **Control**: A few sprays of Borax (0.6%) at fruit set and afterwards are beneficial with the wind ward side of kilns or keeping the height of their chimneys at least 15-18 meters.
- **Malformation:** Reported by Watts in 1891 in Darbhanga in Bihar. It might be physiological, fungal, virus etc.
- **Spongy tissue**: First observed by Cheema and Dhani in 1934, develops in the pulp (Mesocarp) of the fruit during ripening, loss up to 30%.
- **Control**: Two preharvest sprays of 0.1 per cent Bvistin at the time of 10 and 20 days harvesting, post-harvest dipping in 500 ppm Bvistin.

Diseases

- **Anthracnose**: It is caused by *Colletotrichum gloeosporioides* or (*Glomerella cingulata*) fungal disease, prevalent in humid and high rainfall areas. The leaves, shoots inflorescence and the fruits are all affected by it.
- **Control**: Spraying of Bordeaux Mixture (3:3:50), Blitox or Phytolan (0.3%), Bavistin (0.1%) thrice a year i.e. February, April and September.
- **Powdery mildew**: (*Oidium mangiferae*) fungal disease, losses range from 30-90%, it can be controlled by a spray of wettable Sulphur (0.2%) just after the emergence of panicle, Carbendazim (0.1%) and Dinocap (0.1) at 10-15 days interval.
- **Canker**: It is caused by *Xanthomonas campestris* pv. *Mangifera indica*. The spray of Streptocycline (100 ppm) at 10-day interval to control, *Bacillus coagulans* is the most potential biocontrol of bacterial pathogen.

Pests of Mango

- **Mango Hoppers (*Amritodus atkinsoni* and *Idioscopus niveosparsus*)**

  Most serious pest for northern and eastern India. Spray of Carbaryl (Sevin) 0.05%, monocrotophos 0.04%, phosphamidon 0.05% at panicle emergence and peanut stage of fruits.

- **Mealybug – (*Drosicha mangiferae*)**

  It is a serious pest of mango. Nymph and adult females suck the sap from the inflorescence. Use of slippery bands 30-40 cm wide, 30-40 cm above the ground level in the month of Nov-Dec. Spray of Carbaryl (Sevin) 0.2%, monocrotophos (Nuvacron) 0.04%, BHC-0.2% is used to kill the nymph.

- **Fruit fly**: (*Dacus* spp.: *Anastrepha obliqua*, *Bactrocera tryoni*)

  Most of the late varieties of mango are very susceptible to the attack of this insect. Vapour Heat Treatment is used to control the attack for export of mango.

  Spray of Carbary 0.02% + 0.1% hydrolysate, repeat the spray after 21 days. Hang traps containing 100ml emulsion of methyl eugenol 0.1%, malathion 0.1% from April to June in the mango orchard.

- **Stone or nut weevil** (*Cryptorhynchus sternochetus Mangifera*; *Cryptorhynchus gravis*) use of irradiations (Sevin) 0.2%, monocrotophos (Nuvacron) 0.04% to kill the nymph.

- **Hopper** (*Amritodus atkinsoni* and *Idioscopus niveosparsus*)

  Most serious pest of mango in the northern and eastern parts of India. Spray of monocrotophos 0.6% or dimethoate 0.06% at 21-day interval during August-October and in February.

- **Stem borer** (*Batocera rufomaculata*)

  Plugging the hole after pouring kerosene oil/petrol or formalin or 0.5% monocrotophos.

9

Apple

Common Name	-	King of temperate fruits
Botanical Name	-	*Malus* × *domestica* Borkh (*M. sylvestris* × *M. dasyphyllum* × *M. praecox*)
Family	-	Rosaceae
Chromosome No.	-	34, 51 X=17
Climatic adoptability	-	Temperate
Fruit morphology	-	Pome
Rate of respiration	-	Climacteric
Photoperiodic response	-	Long day plant
Relative salt tolerance	-	Highly sensitive
Self-incompatibility	-	Gametophytic
Type of Pollination	-	Cross-pollination
Center of origin	-	Asia minor to western Himalayas
Aroma of fruits		
Ripe	-	Ethyl-2-methyl butyrate
Green	-	Hexanal
Established	-	Central Institute for Temperature Horticulture (CITH) Srinagar J&K-1991
Acid present in fruit crops	-	Malic acid
Area (India)	-	301.04Thousand Ha
Production (India)	-	2326.90Thousand MT
Productivity (India)	-	7.73 MT/ha
Top producer in India	-	Jammu and Kashmir> Himachal Pradesh> Uttarakhand> Arunachal Pradesh> Kerala

Top Productivity in India	-	Jammu and Kashmir (11.43 MT/ha)> Nagaland (7.37 MT/ha)> Himachal Pradesh (3.96 MT/ha)> Uttarakhand (2.32 MT/ha)> Arunachal Pradesh (1.58 MT/ha)
Top Producing in World	-	China> USA> Poland> Turkey>India
Top Productivity in World	-	Chile (48.80 MT/ha)
Export from India	-	Nepal> Bangladesh> Iran> Qatar> United States
Post Harvest Losses	-	10.39
Fruit harvesting period	-	Sept-Oct.
Storage temperature	-	-1.1-0°C
Storage life	-	4-8 months
Propagation method	-	Tongue grafting
Spacing (m2)	-	7.5 m × 7.5 m
Planting Time	-	Jan-Mar
Largest producer in the world	-	China Alternate
Bearing habit	-	Apple
Apomixes	-	Apple
Inflorescence	-	Cymes
Pollination	-	Insects (Entomophilies)
Edible part	-	Fleshy thalamus
Ethylene production	-	High 10-100µLC2H4 /kg/hr. 25-2500µL/L
Respiration rate	-	Low 5-10 mg CO2
Bearing habit	-	Axillary (old season growth)
Fruit bud	-	Mixed bud

- Seeds stratification at 4-7°C for 60-90 days from December to February.
- Scions cannot be used for grafting at their active stage of growth.
- Meadow orcharding: 20,000-70,000 plants/ha (Super high density)
- Bitter pit of apple: Golden delicious, yellow (due to calcium) Newton.
- Internal browning: Newton
- Himachal Pradesh is known as the "Apple bowl of India"
- Flower colour: White to Pink

- Apple is the most widely grown temperate fruit in the world
- India rank 5th in apple production in the world
- Apple crop accounts for 55% of the total area and 75% of the total production of temperate fruits in the country.
- 25% to 33% of pollinating trees in India are recommended for regular cropping.
- San Jose scale was introduced in India from France in 1906.
- Apple cider-fermented wine is prepared from apples.
- During the active growth period, average summer temperature should be around 21-24°C.
- The most critical period for water requirement is during April-August.
- Thomas Andrew Knight produced the 1st apple cultivar named as Thomas Andrew Knight.
- Mixed bud bears terminally occasionally on the shoot and principally on spurs.

  Problems observed: High temperature causes water core of the apple, Excessive water leads to rough bark core in apple, Water deficiency causes Jonathan spot of apple

Rootstock

- Seeding – crab ('Trel') apple: *Malus baccata*–Most commonly used rootstock in India.
- M.9 dwarf rootstock suitable for HDP
- M.4, M-7 and MM-106- Semi-dwarf–suitable for HDP
- MM.111 Semi-vigorous: Resistant to woolly aphids and drought
- Merton 793- vigorous: Resistant to woolly aphid and suitable for replant problem.
- M.27 (M.13 × M.9) Ultra dwarf for HDP
- MM.104: Most winter hardy clonal rootstock
- Others – Bud.9, Bud.146, P.22 & Ottawa 3

Early –varieties

- Tydeman's early, Irish peach, Fenny early, Shanburry, Benoni.
- Mid – Red Delicious, Ric-a-Red, Top Red, Red Chief, Red Gold, Jonathan, Rome Beauty, Razakwar, McIntosh, Cortland, Golden Delicious

- Late – Yellow Newton, Winter Banana, Granny Smith, Lal Ambri, Rymer, Buckingham, Pink Lady
- Akbar – (Ambri × Cox Orange Pippins)
- Green English Varieties – Baldwin, Cox's Orange Pippins, Black Bendavis, Pippins
- Spur Type Varieties – Stark Crimson, Silver Spur, Oregon Spur, Well Spur, Red Chief, Red Spur, Super Chief, Hardi Spur, Scarlet Spur, Ace Spur
- Standard Colour Mutant Varieties – Top Red, Skyline, Hardiman, Supreme to Vance Delicious, Bright-N-Early, Kalpa-1 to Starking Delicious
- Pollinizer Varieties: Golden Delicious, Red Gold, Tydeman's Early, Worcester, Summer Queen, Golden Spur, Granny Smith, Winter Banana, Mcintosh, Gala and Fuji flowering crops like Manchurian, Snowdrift and *Malus floribunda*
- Scab Resistant Varieties – Prima, Priscilla, Sir Prize, Jonafree, Florina, Mecfree Nova Mac, Red Free, Nova Mac, Liberty, Freedom, Firdous Shireen, Florina (France)
- Low-Chilling Varieties – Anna, Dorset Golden, Michel, Tamma, Tropical Beauty, (Parlien's Beauty Suitable for processing)
- Crab Apple – Red Flush, Crimson Gold Yellow Drops, Snowdrift
- Good Seed Viability – Golden Delicious, Yellow Newton, Northern Spy Mcintosh.
- Delicious group of varieties are self-incompatible and require cross-pollination
- English group of varieties are self-pollinated and act as a suitable pollinizer for the delicious group.
- The Red Delicious is the most popular variety of India. It was introduced in Himachal Pradesh by Satyananad Stokes around 1917 and its cultivation gradually spread to Kashmir and Uttarakhand
- Red Gold – acts as a pollinizer for Red Delicious and Starking Delicious.
- The Golden Delicious – is renowned as the Commercial Cavity of the USA and Europe.
- McIntosh is the leading variety of Canada.
- The Ambri-Indigenous variety has the longest storage life and is grown in Kashmir. .
- Rymer is an indigenous variety grown in Kashmir.
- Triploid cultivars – Baldwin, Gravenstein, Winesap

- Northern Spy is resistant to woolly apple aphid.
- Ambri has long keeping quality but is biennial in bearing.
- Liberty variety of Apple is resistant to all fungal diseases.
- Gala and Fuji: High yielding strains.
- Gala, Fuji, Golden Delicious and Red Gold are recommended pollinizers for apple.
    - Lal Ambri – Red Delicious × Ambri mainly growth in Kashmir =-
    - Amb Red – Red Delicious × Ambri
    - Amb Starking – Starking Delicious × Ambri
    - Amb Royol – Royal Delicious × Ambri
    - Amrish – Rich-a-Red × Ambri
    - Sunheri –Ambri × Golden Delicious
    - Chaubattia Princess – Red Delicious × Early Shanburry
    - Chaubattia Anupam – Red Delicious ×Early Shanburry
- Ambred, Ambstarking, Ambroyal, Ambrich have been developed at (RHRS) Mashobra H.P.
- Lal Ambri, Sunehari – (IFRS) Shalimar (Jammu and Kashmir) was developed in 1956.
- Chaubattia Princess and Chaubattia Anupam were developed by Horticulture Experiment and Training Centre, Chaubattia, Uttarakhand
- The Red Delicious cultivar is considered as the mother of all delicious groups of the cultivars.
- Most of the commercial apple varieties are diploid; while triploids are rare.
- The diploids are self-fruitful and triploids are self unfruitful.
- Malling Merton (MM) series of rootstock is the result of a cross between Malling (M) series with woolly apple aphid resistant variety of Northern spy. These were developed by East Malling Research Station, Kent in Collaboration with John Inns Irrigation Institute, Washington, England
- EMLA, East Malling Long Ashton series of rootstock is resistant to viruses.
- Bitter pit in apple is caused due to Ca deficiency.
- *M. Sikkimiensis*, *M. Torngoides* & *M. hupehensis* are apomictic rootstocks of apple.
- Resistance to powdery mildew in cultivated apple is transferred from *M. zumi.*

- *Aphelinus mali* which is a predator of woolly apple aphid.
- *Malus* genus has 25 species.
- Apple was introduced in India by Captain Lee in 1865 in Kullu Valley of H.P.
- Later on, red coloured delicious group varieties were introduced at Kotgarh in Shimla (H.P) in 1917 by American Missionary Mr. Satyanand Stokes.
- Apple fruit is rich in carbohydrate (15%) protein (0.3%) and nutrients viz. K, P and Ca.
- Low chill varieties require only 200-800 chilling hours.
- The optimum temperature for pollination, pollen germination and fruit setting is 18 to 22°C.
- Well distributed rainfall of about 100-125 cm throughout the season is considered most favourable.
- Soil pH – 6.0-5.5.
- The stratified seeds are sown in nursery beds during March at a spacing of 8-10 cm from seed to seed and 15cm-20 cm from line to line.
- The seedling rootstock attains graftable size of 15mm diameter in a year.
- The clonal rootstock is commercially propagated through mould layering or trench layering.
- The best time for grafting of apple is during February-March with tongue and cleft methods.
- Chip budding can be also done in March.

Spacing and planting density for different scion stock combinations:

Scion	Rootstock	Tree Size	Spacing (m)	Tree/he
Standard	Seedling	Vigorous	7.5 × 7.5	178
Standard	MM.111, M.793	S. Vigorous	6.0 × 6.0	278
Standard	MM.106, M.7	S. Dwarf	4.5 × 4.5	494
Standard	M.9, M.26	Dwarf	1.5 × 1.5	4444
Spur type	Seedling	S. Vigorous	5.0 × 5.0	400
Spur type	MM.111, M.793	S. Dwarf	3.5 × 3.5	816
Spur type	MM.106, M.7	Dwarf	3.0 × 3.0	1111

- The objective of pruning is to maintain a proper balance between vegetative growth and spur development.
- Training is completed during the initial 4-5 years after planting.

- The best time of pruning is during dormant season (January-February)
- The initiation of flower primordial starts about 3-6 weeks after full bloom (June).
- The influence is determinate having five flowers. Flowers are white, pink or carmine in cymes.
- Flowers of most cultivars are epigenous and hermaphrodite.
- The flower consists of five petals, five sepals, 15-20 stamens and a pistil which is divided into five carpels each containing two ovules.
- Sterility and incompatibility are the two main causes of unfruitfulness in apple.
- In the orchard of optimum fertility, NPK is applied in the ratio of 70:35:70g/ year age of apple tree and doses of these NPK fertilizers are stabilized at the age of 10 years.
- For ten or more than ten-year old apple tree; 100kg FYM. NPK-700:350:750g should be given annually.
- FYM along with full dose of PK and half (1/2) dose of N are applied during December-January.
- Half (1/2) dose of nitrogen is applied in two split doses. Half of half (1/4) dose of N is applied one month before flowering (March) and the remaining half (1/4) dose one month before fruit harvesting.
- It can be corrected by two foliar sprays of ZnSO4 (0.5%) and boric acid (0.01%) before flowering and one more during May-June.
- The intercrops like peas, beans, cabbages, cauliflower and ginger can be cultivated in the initial years of plantation
- For herbicidal weed control-Simazine-4kg/ha, and Glyphosate @8kg/ha – July and August.
- The most critical period of water requirement is from April to July when flowering, fruit setting, fruit growth and development occur.
- Fruit thinning by foliar spray of 20ppm NAA at petal fall results in optimum thinning.
- For treatment of Pre-harvest drops; control of 10ppm NAA 20-25 day before harvest is effective.
- Breaking seed dormancy – 100-200ppm GA3
- Rooting in cutting – IBA – 2000 ppm and NAA@200 ppm
- Application of promoting 30-60ppm GA_4+7+Cytokinin at pea stage helps to improve the shape of apple fruits.

- Application of ethereal @1200ppm or 4.5ml ethereal/litre of water + 25ppm NAA improve fruit quality.

Maturity indices for important varieties of apple.

Variety	DFFB (Days)	Firmness (kg/cm2)	TSS%
Tydeman's Early Worcester	90±4	7.8±0.15	12.0-13.0
Starking Delicious	120±5	8.2±0.4	13-15
Red Chief	110±5	8.5±0.4	14-15
Red Delicious	134±5	8.5±0.4	13-14
Golden Delicious	148±5	8.4±0.4	12-14.5
Granny Smith	180±5	8.5±4	130-14

- The usual dimensions of telescopic CFB with trays are 50.4 × 30.3 × 28.2cm outer jacket 50 × 30 × 28.2cm (inner case).
- Flowers of most cultivars are epigenous and hermaphrodite while the ovary is inferior.
- Pollinizer-25-33%, placement of 5-6 honey bees colonies per ha.
- Top working of 2-4 shoots of commercial varieties with pollinizers.

25% Pollinizer	33% Pollinizers
O O O O O O O	O X O X O X O
O X O X O X O	O X O X O X O
O O O O O O O	O X O X O X O
O X O X O X O	O X O X O X O
O O O O O O O	O X O X O X O
O X O X O X O	O X O X O X O
O O O O O O O	O X O X O X O
O X O X O X O	O X O X O X O
O O O O O O O	O X O X O X O
O X O X O X O	O X O X O X O

O = Commercial variety, X= Pollizer varieties

- The fertilizer should be broadcasted in the basins 30cm away from the trunk.
- Fruit drop: 40-60% fruit drop in three phases
- Early drop – lack of pollination
- June drop – Moisture stress
- Pre-harvest drop – reduction in levels of auxin and increase in ethylene in the fruit.
- Application of 10ppm NAA 20-25 days before harvest checks this drop.

Use of growth regulations

- Plant propagation
    - Seed dormancy: GA3 at 100-200ppm
    - Cutting: IBA – 2000ppm
    - NAA-200ppm
- Effect on growth: Auxin and GA3 help in increasing vegetative growth
- Application of 250ppm GA3 stimulates vegetative growth besides also enhances alternate bearing phenomenon.
- Effect on fruit set and yield: Fruit setting Triacontanol – 20ppm
    - Miraculan – 0.6ml/L
    - Paras – 0.6ml/L
    - Biozyme – 2ml/L

  Sprayed at bud swell and petal fall stages helps in improving fruit set and yield.
- Fruit drop: treatment with 10ppm NAA, (10ml of Planofix in 4.5L of water) one week before the expected fruit drop.
- Fruit thinning: 10-20ppm NAA, 7-10 days after petal fall.
- Improvement of fruit shape: Application of Promalin 30-60ppm (GA3+7+Cylokinin) at pea stage helps to improve the shape of apple fruits.
- Improvement of fruit colour and maturity: Application of Ethereal 1200ppm or 4.5ml Ethrel / litre of water + 25ppm NAA improve surface and colour in apple.
- About 30% of post-harvest losses have been recorded which can be minimized by proper handling of fruits.
- Pre-cooling: Immediately after picking, the fruit should be placed in a cool and ventilated place for removal of field heat before packing.
- For the removal of field heat, several methods such as use of air cooler, cold water sprinkling, fruit washing and keeping of fruit overnight in a cool place are adopted.
- Storage for 4-6 months -1.1-0°C with 80-90% RH.
- Processing: About 30% of apple fruits are rejected as cull fruits before packaging for the fresh fruit market.

  Products like Jam, Fruit Leather, Fruit toffees, wine can be made from fresh fruit juice.

Hybrid Apple Cultivars

- Vance Delicious – Mutant of Delicious
- Top Red – Bud mutant of Shotwell Delicious/Starking delicious
- Kalpa-1- Bud sport of Starking delicious
- Red Chief – Limbs Sprout of Star krimson.
- McIntosh – Chance seedling
- Mollies Delicious – Golden Delicious × Edge Wood × Gravenstein
  Fuji – Rod Delicious × Rolls Janet) 1958 Named as Fiji in 1962 in Japan 51% production.
- Gala – Kidds Orange × Golden Delicious 1939
- Jonagold – Golden Delicious × Jonathan
- Empire – McIntosh × Delicious
- Criterion – Starking Delicious × Golden Delicious

Rootstocks of Apple

- M27 – East Malling UK (M13 × M9)
- B146 –Michirinsk College Russia
- BM427 – Balsgard Sweden (M4 × Antonoka Kamensischaka)
- G65 – Cornell University (M27 × Beauty crab apple) New York the USA
- JM1,5& 8 – Apple Research Center (Marabakaida × M9) NIFTS Japan
- J-TE-G – Techobeezee Czech (Max Croncels) Republic
- P22 – Skierwiowice Poland (Max Antomovoka)
- Vincent Canada
- V3 – Vindant Canada
- Voinesti 2 – Voinesti Romania (Max Cretesc)
- MAC9 – Midigan Stata University the USA
- Ottawa 3 – Ottawa Canada
- Denali – Balsgard Sweden
- KSC-28 – Kentuille Nava Scotia Canada
- MM – East Malling UK
- G30 – Geneva Research Station, New York USA (Robusks X Ma)
- Meston – John Inner Institute UK (Max Northern spy)

- Alnasp – Alinasp Research Station Sweden
- Nicole – Geneva Research Station New York

Origen of apple rootstocks

- United States : CG, MAC, OAR, Ark and MN series
- Canada : KSC, Ottawa, Vineland series
- United Kingdom : M, MM, and AR series
- Germany : J9 and Pillnitzer-Supporter series
- Sweeden : Alnarp, BM series and Bemali
- Russia : Budagovskij series
- Poland : P series
- Czech Republic : JT-E series
- Israel : MH series
- Romania : Voinesti series
- Japan : JM series

The training system in apple

- Central leader system was developed by Heinrike (1975) in North America and Mckenzie (1972) in New Zealand
- Spindle bush developed at Germany in 1940
- Hybrid Tree Cone Orchard System (HYTEC) developed in late 1980 by Barritt (1991-92)
- Slender Spindle – Holand 1915

San Jose Scale (*Quadraspidiotus perniciosus*)

- By lifting armature, a yellow-coloured insect under-earth is seen. The nymph and adults suck sap from aerial parts.
- Spray 2% dormant oil (Servo orchard oil / Hindustan Petroleum spray soil) or 1.5% summer oil like orchaks 796/1POL/ shelter 909 at half leaf to light cluster stage.
- 0.04% Chlorpyrifos (200ml Durmet in 100 L water)

Woolly apple aphid (*Eriosoma lanigerum*)

- It feeds on apple fruit and lives in colonies on the aerial parts and roots of plants.

- Damage is caused by sucking of sap from stem, twigs and roots resulting in gall formation, plants remain stunted.
- 0.04% Chlorpyrifos (200ml Durmet in 100L water)
- During May-June and again in October. (infested tree)
- 0.1% Chlorpyrifos in October-November using 10-15 litre solution (infestation on roots).

European Red Mite (*Panonychus ulmi*)

- The maximum population is observed during May-July. The mite completes 5-7 generations in a year.
- Spray 2% dormant oil and 1% summer oils.
- If its population is high, spray with Fenazaquin (25ml magister 10EC/100L) of Propargite (100ml omite/100L water) twice at 20 days interval in June-July.

Apple Scab (*Venturia inaequalis*)

- Light brown or olive green spots which soon turn musty black appear or either or both sides of the growing leaves in spring.
- Small lesions develop on the fruits, and increase gradually leading to mishappening and cracking of fruits.
- Spray schedule of Dodine (0.1%), Mancozeb (0.3%)

  Mancozeb (0.3% + Carbendazim (0.05%) at pink buds Benomyl / Carbendazim (0.05%) at Patel fall- Zineb (0.3%).

  Urea 5% spray after fruit harvest to control the disease

Powdery mildew: (*Podosphaera leucotricha*)

- Mycelium leaves on dormant buds
- Wettable sulphur (200-300g/100L),
- Contaf (50g/100L), Baycor (50g/100L)

Canker

The spray of Copper oxychloride (300g) in 100L water

Captain (200g) in 100L water

White root rot: (*Dematophora matrix*) fungi

- Copper oxychloride (300g/100L) Nov-Dec on cut ends of the roots.

Carbendazim (100g) along with Mancozeb (300g) in 100L of water during April, June, July and September in infected trees.

- Orchards more than 40 years of age face the problem of unfruitfulness more seriously than the young orchards.
- Use of Ethephon helps attaining maturity and improves colour at lower elevation.

Premature leaf fall: (June-July)

- The disease was first noticed in 1995 in H.P.
- All the commercial Delicious cultivars are susceptible.
- Causal organism : *Marssonina coronaria*

  Symptoms first appear as dark green circular patches on the upper surface on the mature leaves giving rise to 5-10 mm size brow leaf sports.

  Management: Proper orchard management, Urea-5%, Protective 3-4 sprays of fungicides like Mancozeb 0.3%, Carbendazim – 0.05%, Thiophanate methyl - 0.1%, Benomyl – 0.05%, Propineb – 0.03%, Dithianon – 0.075%, Ziram – 0.3%, Dithianon – 0.05%, Zineb – 0.3%.
- Modified central leader system is most suitable for standard plantation.
- MCL – Headed back 60-70cm above the ground level.
- The best time of pruning is during the dormant season i.e. Dec-Jan and summer season i.e. April-May.

Causes of low productivity of apples

1. **Climatic factors**: Low temperature at the time of flowering and fruit set. It is a well established fact that the flowers die below 22•C and bee activity is completely inhibited below 4.4°C.
2. **Varietals factors**: In Himachal Pradesh, the Delicious group of apples occupy more than 80% of the entire cultivation which is self-unfruitful and require cross-pollination for fruitfulness.
3. **Inadequate pollinizer**: Pollinizer required – 25-33%.
4. **Lack of pollinators**: Honey bees are the major agents.
5. **Inadequate nutrition**: Frequent dry spells during April-June and September-November make the nutrients unavailable to the plants even if applied adequately in the soil.
6. **Poor soil conditions:** adequate drainage, temporary waterlogging conditions.

7. **Poor canopy management:** in such conditions lesser heading back and more thinning out of shoots as per the behaviour is required to balance cropping and growth.
8. **Senile orchard:** Needs rejuvenation or replanting
9. **Pathological factors:** Adoption of recommended POP required
10. Entomological factors:

Aphid	82%
San Jose Scale	71%
Blossom Thrips	70%
European mite	62%
Apple leaf roller	43%
Root borer	26%
Stem borer	9%
Defoliating beetles	6%
Hairy caterpillar	5%

Replant Problem

Causes

- Many fungi belonging to oomycetes, hyphomycetes and basidiomycetes have been reported to be the causal agents of replant disease.
- Many investigators In England, Canada and USA have shown that species of *Phytophthora* and *Pythium* are the primary cause of replant disease. *Pythium syluastium* in Canada.
- In Himachal Pradesh, India, we have identified the potent fungal pathogens (*Dematophora necatrix*, *Fusarium oxysporum*, *Phytophthora cactorum, Pythium ultimum* and *Rhizoctoina solani*) and nematodes (*Tylenchorhynchus mashhoodi*, *Pratylenchus coffeae*, *Xiphinema* spp., *Paratylenchus curvitatus* and *Helicotylenchus dihystera*) causing the apple replant problem.

Management

- This problem can be controlled at replanting orchard sites by integrated management practices. The soil fumigation and solarization have been found somewhat more effective, but due to the complex nature of the problem, the combined use of chemical, cultural and biological methods could be beneficial in controlling the disease. Use of certain resistant clonal rootstocks and selection of right scion and rootstock combination as per the agro-climatic conditions may also be helpful in controlling this problem.

- Inter-cropping: with mustard, radish, and antagonistic crops, such as marigold (*Tagetes patula*) successfully reduced the population of nematodes, *Pratylenchus penetrans* and *Pythium* spp.

10

Banana

Botanical Name	-	*Musa paradisica* (Monocotyledoneus)
Common Name	-	Sanskrit: Moka, Bible: Tree of wisdom, Koran : Tree of Paradise, Kalpataru, Apple of paradise, Adom's fig
Family	-	Musaceae
Chromosome No.	-	(2n) 22, 33, 44. X=11
Climatic adaptability	-	Tropical
Fruit morphology	-	Berry
Rate of respiration	-	Climacteric
Photoperiodic responses	-	Lon day plant
Type of dichogamy	-	Protogyny
Pollination	-	Bats and Birds
Centre of origin	-	South East Asia
Aroma	-	Green - Hexanal Ripe - Eugenol Over ripe - Isopentanol
Acid	-	Malic acid
Maximum area	-	Karnataka
Maximum production	-	Andhra Pradesh
Maximum productivity	-	Punjab
Storage tem.	-	12.8°C (13°C)
Storage life	-	3 weeks
Harvesting period	-	3-5 months after flowering
Propagation method	-	Sword sucker (700-1000g)
Planting spacing	-	1.8 m × 1.8 m
Planting time	-	February-March; September - October
Optimum temp for growth	-	20-30°C
Maximum producer in the World	-	India

Area - 883.77 Thousand Ha.

Production - 30807.50 Thousand MT.

Productivity - 34.86 MT/ha

Maximum productivity in the World - Indonesia (50.6 MT/ha)

Parthenocarpy - Vegetative

Inflorescence - Spadix

Pollination - Birds (cross-pollination)

Edible part - Mesocarp and Endocarp

Ethylene production - Medium (1-10) µl C_2H_4/kg/hr

Respiration rate - Medium (10-20mg) Release of CO_2

Bearing habit - Terminal (Old season growth) (Unbranched)

Flower bud differentiation - September-April

Established - National Research Centre for Banana Tiruchirapalli, Tamil Nadu-1993

Sword sucker - Before planting dipping in 0.5% Monocrotophos and 0.2% Bavistin

Drip irrigation - 40-45% saves water and early harvesting

High density planting - 3000-4000 plants/ha

Double row system - Planting is adopted in Maharashtra and Gujarat

Bunch covering - With polythene in November increases yield up to 25-30%

KH_2PO_4 - Two sprays at the fruit development stage increase the bunch weight

Top Producers (World) - India> China> Indonesia> Ecuador> Brazil> Philippines

Top Producers (India) - Andhra Pradesh> Gujarat> Maharashtra> Tamil Nadu> Uttar Pradesh >Karnataka

India's contribution - 15% of the total world production

Flower - Female and hermaphrodite

- It is controlled to parthenocarpy and sterility which can be achieved by triploids. Wild species - *M. acuminata* and *M. balbisiana*
- The cultivated edible banana is mainly triploid.
- The major problems in the breeding of banana are sterility, polyploidy and parthenocarpy.

Introduction cultivars

Cultivar	Introduced from	Characters
Ladyfinger	Australia	Resistance to bunchy top
Grand Nain MS	France	by D.R tissue cultured plant

First breeding globally	-	First started at (ICTA) Imperial College of Tropical Agriculture in Trinidad (1922) and later in Jamaica in 1924
First breeding in India	-	In 1949 at Central Banana Research Station Aduthurai, Tamil Nadu

- The underground portion of banana is a modified form of stem known as rhizome.
- The portion above the ground made of sheaths of leaves is known as pseudostem.
- Banana initiates flowering 9-12 months after planting.
- Sword sucker with narrow, slender leaf blades is the best planting material.
- Banana is a moisture and nutrient loving shallow-rooted plant.
- Simmonds and Shephard (1955) devised a scoring technique for genomic classification of banana.
- Pollination in banana is mediated mainly by bees and birds.
- Banana fruit follows the double sigmoid growth curve (seeded -sigmoid parthenocarpic - double sigmoid)
- The basic chromosome number of Essmusa and Rhodochlamys section of *Musa* is 11, *Callimusa* is 10, *Musa ingens* is 7 and *Musa beccarii* is 9.
- There are six classes of genomic classification: two diploids, three triploids and one tetraploid.
- *Musa ingens* is the largest known herb which grows upto the height of 10 m.
- Harichal (AAA) is a semi-tall sport of Dwarf Cavendish.
- Frost is the major limiting factor of banana cultivation. Banana is a crop of humid tropics.
- Waterlogging condition leads to incidence of Panama wilt in banana.
- Desuckering (removal of unwanted suckers), propping (support given to fruit bunch bearing plant) and mattocking (sequential cutting of the plant after harvest) are important practices in banana cultivation.
- Potassium deficiency in banana results in improper bunch filling.

- Occurrence of fingertip disease is severe in high-density plantations
- Zinc deficiency in banana causes bunchy top crowns, narrow pointed and chlorotic young leaves.
- Application of N at reproductive phase delays senescence of leaves and improves bunch weight.
- Vegetative phase (1-6 months) is crucial for weed control in banana weeds causing 60-70% yield loss.
- Bunch covering with polythene and 2 sprays of KH2PO4 at fruit development stage increases fruit yield in banana.
- Nematodes management in banana induces double paring and treating the sucker with 0.5% monocrotophos, growing of sunhemp as in intercrop. Cropping system of banana-rice-banana is beneficial for nematodes control.
- Tetrazolium test is used for the diagnosis of presence of Banana bunchy top virus.
- Banana Streak Virus (BSV) is a serious disease transmitted through mealybug (*Planococcus citri*).
- Banana Bract Mosaic Virus (BBMV) is a serious infection in Monthan banana.
- Burrowing nematode (*Radopholus similis*) is commonly present in banana plantations.
- Pisang Lilin and Fongat are resistant to the nematode.
- Gros Michel, though a high yielding variety of banana is susceptible to Panama wilt.
- Robusta banana is highly susceptible to bunchy top, anthracnose, cigar end rot, blacktip and frost but resistant to Panama wilt.
- Bunchy Top, the viral disease is transmitted through banana aphid (*Pentalonia nigronervosa*)
- 'Moongli' a mutant of Nendran has no male phase and plants bear one or two hands with the biggest fruit.
- Seeded banana follows a single sigmoid growth curve while seedless exhibits double sigmoid growth curve.
- Flower visiting insects are the main agents for transmitting the Moko disease in banana.
- Seediness in banana can be controlled by a spray of 2,4-D @25ppm within a week of opening of the last hand.
- Bunchy top of banana was first observed in 1891 in Fiji.

- Ripe banana contains 27% sugar.
- In 'Monthan' variety of banana, only glucose sugar is found.
- *Musa acuminata* is the source of today's edible banana.
- 'Gross Michael' banana is susceptible to Panama wilt while 'Basarai' is immune and 'Poovan' is resistant to this disease.
- The bunchy top is also called cabbage top.
- Male flower of banana is resistant to panama wilt disease but susceptible to bunchy top disease.
- India's share in world's banana production is 25.70%.
- Banana is a calcifuges crop (Calorific value – 67-137/100g)
- Sigatoka is a serious disease of banana.
- Removal of male bud after completion of the female phase is referred to as 'denavelling'.
- In Tamil Nadu, banana is grown especially for leaf production.
- Fingertip disease is serious in high-density planting.
- Genetic classification of banana was formulated by Simmonds and Shepherd.
- The temperatures above 36-38°C having scorching effect with increased transpiration rate.
- Besides sword suckers (weighing 500-750g), cut rhizomes called 'Bits' and 'peepers' are also used for propagation.
- Poovan, Rasthali, Mendran and Robusta require 2.1 m × 2.1 m spacing.
- Basari, Kulhan, Jawari require – 1.8 m × 1.8 m spacing.
- Method of planting: Furrow method is used in Gujarat and Maharashtra; and Trench method is popular in Tamil Nadu:
- AAB and AAA clones are grown under irrigated conditions.
- ABB clones are grown under rainfed condition (Monthan, Kanthali, Kunnan).
- Trench method is especially used in the wetland system of cultivation.
- Desuckering once in 45 days is a common practice in the banana plantation.
- For getting maximum yield, minimum of 10-12 leaves are required to be retained on the mother plant.
- Strong wind is a threat to successful banana production.
- For long-distance transportation, harvesting is done at 75-80% maturity.
- Skin coating with waxol (12%) wax emulation is a post-harvest technology employed to delay ripening of banana is.

- All AAA clone are susceptible to Sigatoka leaf spot.
- Banana cultivars are screened for viruses using ELISA test.
- Singtoka leaf spot disease of banana was first observed in 1913.
- Poovan, Monthan, Karpuravalli groups are some of the high yielding varieties:
- Banana is the staple food of South Africa.

Chief Varieties and their characteristics

Dwarf Cavendish (AAA)	- Basrai - leading commercial cultivar contributing 58% to the total production. Yandevi selection (Hanuman or Pardase) from Basrai.
Robusta (AAA)	- Bombay green and Harichal - the semi-tall sport of dwarf Cavendish, highly susceptible to Sigatoka leaf spot but resistant to Panama wilt.
Grand Naine (AAA)	- Full mutant of dwarf Cavendish. It requires propping.
Poovan (AAB)	- Rasthali, Amritpani, Mortman (choicest of table bananas) Hand lumps (cracking and physiological disorder). Tolerant to many abiotic & biotic stresses.
Poovan mysor (AAB)	- Pink pigmentation on the ventral side of midrib when young, susceptible banana streak virus, leading culture of South India.
Nendran (AAB)	- Rajeli is the most prized cooking variety grown in Kerala using trench plantation. It is good for making banana chips.
Hill banana (AAB)	- Virupakshi, Sirumalai, Haden-suitable for cultivation on hills. Fruits have unique aroma and flavour (taste) making them suitable for Jam making.
Lal Velchi (AAA)	- It is grown for red skin.
Manthan (ABB)	- Good for culinary purpose.
Ney Poowan (AB)	- Diploid variety that fetches double price than other cultivars due to horizontal bench orientation.
Pey Kunnan (ABB)	- 'Karpuravalli' and 'Kanthali' -tolerant to biotic and abiotic stresses. Good for baby food, juice, wine and popular in marginal soils.
	'Amritsagar', 'Dudhsagar', 'Chakia', 'Monohar'

Ladyfinger (AB)	-	Diploid banana variety plants are resistant to Panama disease but susceptible to leaf spot and poor yielder. The fruits are believed to have medicinal value.

Hybrids varieties

FHIA-I (Gold finger)	-	Belong to pome group. Resistant to wilt and sigatoka leaf pot.
Bodles Altofort	-	(AAA) synthetic hybrids are a result of cross between Gross Michel (AAA) Pisanglin (AA).
Klue teparod	-	(AABB) natural tetrapoid
Co-1	-	'Kellar lander' × *Musa balasiana* × 'Kadali' developed in 1984

- Poovan & Ney Poovan are preferred in the multistory system.
- Saltwater treatment is used to reduce the duration of banana fruits.
- Banana is a rich source of dietary potassium (K) used in nervous impulses and a good source of energy.

Rajapuri	-	Resistant to cold
Moongli	-	Mutant of Nendran
Edible part	-	Starchy parenchyma

- Brinjal and cucurbits should not be grown in the banana orchard because they attract nematodes.
- Simmonds (1962) opined that table banana might have originated in Malaysia whereas the cooking banana or the *Bulbisian* species originated in Southern India.
- The sheath also produces low grades fibre for agricultural use.
- Good ornamental plants for the garden are *Musa ornate*, *Musa flauiflora*, *Musa velvtina* etc.
- Banana is a monocotyledonous, perennial, herbaceous succulent plant, propagated through rhizomes, which are sympodial in nature.
- The basal internodes bear female flowers while the epical internodes bear male flowers.
- Diploid (AA), triploid (AAA) and tetraploid (AAAA) arising from the species to *acuminata* and hybrid diploids (AB), triploids (AAB and BBA) and tetraploid (AAAB and ABBB).

- In India, F1 hybrids (Agniswar × Pisang Lilin) resistant to sigatoka disease have been developed.

  M. acuminata (2A=22) × *M. balbisiana* (2B=22)

  Natural ↓ hybridization

  AB (2n=22)

  M acuminata (4B=44) × Tetraploids (2A=22)

  ↓

  Triploids (AAB)

  Tetraploids (4A=44) × *M. balbisiana* (2B=22)

  ↓

  Triploids (AAB)

  Triploids (3A=33) in natural mutation
- The Hindus consider banana as a symbol of divinity and prosperity.
- The Christians regard the banana plant to have heavenly origin.
- The Muslims believe that banana is the tree of wisdom.
- The Buddhists also refer the plant as having divine significance.
- The species under the genes *Ensete* are found in Western Africa and the adjoining Islands.
- India has more than 300 cultivated varieties.

Cooking Banana (AAB)

a) Bontha

b) Sail Cola

c) Klue Teparod

- Rainfall, as stated above, should be ideally 100 mm per month. Higher rainfall may induce disease infections.
- Sucker of 80cm in girth is considered to be the best planting material.
- Best planting time in the Northern hemisphere – July to September.
- Best planting time in the Southern hemisphere – February to December.
- Banana is the first solid food given to babies.
- The unripe fruit is even richer in vitamins, which are not lost even after cooking at a temperature up to 60°C.
- **Nendran** variety contains over 26 per cent sugars. It is used to prepare salted banana chips.

- **Poovan**: The bunch is large weighing about 25 kg and comprises 220 fingers.
- Soil – It can grow well in slightly alkaline soil, but saline soils with salinity exceeding 0.5 % are unsuitable. In alkaline soils, wilt disease is less prevalent.
- Banana can be grown well in a pH range of 6.5-7.5. Alluvial and volcanic soils are the best for banana cultivation.
- The coastal sandy loams as well as the red lateritic soils, the hilly tracts of Kerala are congenial for extracting a good crop of banana.
- **Climate** – the crop thrives in warm and humid tropical climate
- **Temperature** – A temperature range of 20-25°C is ideal for banana production. The growth of banana is retarded at temperatures below 20°C and above35°C.
- The optimal range of 22° to 31°C throughout the year is favorable for cultivation. Rise in the day/night temperatures from 17/10°C to 33/26°C increase the efficiency of absorption of N, P, K, Ca, Mg, Na, Cl, B, Mn, Zn and Cu.
- The phenomenon called "Choke throat" is a major problem with dwarf banana at high altitude banana-growing areas.
- **Light** – Leaf increases as photoperiod from 10-14 hours. In banana all light flux density of 800-1000 µ mol m-2.
- **Rainfall** – Production of banana needs 25 mm of water per week. Average annual rainfall of 200-2500 cm evenly distributed throughout the year is considered favourable.
- **Wind**- Wind velocity of more the 50 km/h causes serious damage to banana plantation. Banana is grown up to an altitude of 2000m above mean sea level.
- **Tissue culture** – Swamy *et al.* (1983) explored the possibilities of banana propagated through tissue culture.

Advantages of using 'in vitro' plantlets

i) Availability of plants in large number within a short time.

ii) Replacement is rarely needed after field planted.

iii) Uniformity in growth and harvesting time.

iv) Produces heavier bunches than conventional sucker.

v) In vitro plantlets are free from nematodes, fungal and bacterial pathogens.

vi) In vitro methods ensure rapid multiplication of new cultivars, horticulturally superior clones etc., within a short period (1:1000 rations per year).

Disadvantages of using 'in vitro' plantlets

i) In vitro plant is costlier than a normal sucker and as such initial establishment cost is much higher.

ii) In vitro plants need extra attention and care after planting since they are sensitive to stress.

iii) In vitro propagation is possible in all types of plants due to soma clonal variation.

iv) There is a risk of virus transmission through in vitro propagation.

v) Mother plants should, therefore, be induced with a monoclonal antibody or DNA probe test to confirm freedom from viruses.

- It has also been seen that micro propagated plants show uniformity during flowering and harvest.

Planting time

- September-October: Kerala, T.N.
- February-March: Hilly slopes of South India
- April, June, August or October: Maharashtra
- Planting density – A spacing of 9 feet (2.7 m) × 10 feet (3.0 m) for tall cultivars and 6 feet (1.8 m) × 6 feet (1.8 m) for dwarf cultivars
- Maximum profit can be extracted from cultivation of Dwarf Cavendish banana when spaced at 2.0 m × 2.0 m or 2.5 m × 2.5 m Closer spacing also produces higher yield of fruits.
- Banana requires 40-50 irrigations from the time of planting to harvesting.
- A water-saving of 24.20 % and 13.40 % increase in yield can be achieved under drip irrigation. Saving nitrogen up to 40 %.
- The deficiency of K also reduces total foliage area, number of hands and fingers on a bunch.
- Magnesium deficiency effects on Mg/Ca ratio of the soil.

  Substitute 75 % by organic sources

  N - 100-250g/plant,

  P - 50-100g/plant

  K - 200-350g/plant
- Foliar spray of urea (4%) in the morning humid hours gives immediate relief form nitrogen scarcity.

- The cultivation cost would be Rs. 1, 50, 000/ha with net income of Rs. 2, 00, 000/ha.
- Application of Glyphosate 2kg/ha or Gramoxone 1.8kg/ha can be used for controlling of weeds.
- The studies on Black polyethylene, sugarcane trash, banana trash mulching and un-mulched conditions recorded 112.9, 95.5, 85.6 and 76.8 t/ha yield, respectively. The cost / benefit ratio was higher in sugarcane trash mulching followed by banana trash.
- **Intercropping** – Radish, cauliflower, cabbage, spinach, chilli, brinjal, colocasia, yam, lady's finger, cucurbitaceous vegetable, marigold, and tuberose can be grown in intercropping.
- Suckering with a portion of corn at an interval of 5-6 weeks hastened and increased the yield by 12 per cent.
- **Earthing up** – During summer and winter, the plants should be grown in furrows and on ridge during the rainy season.
- **Propping** – Propping is also practiced with bamboos. the method of propping cables is adopted in Israel.
- A minimum of 12 leaves requires to be retained for maximum yield.
- There are two types of suckers which originate from rhizome namely sword suckered and water suckers.
- **Sword suckers** – have a strong connection with the mother plant leaves, are small, thin and sword-like and usually originate from the auxiliary bud lower down on the parent rhizome.
- **Water suckers** – usually develop from shallow buds near the soil surface or even above it, because of weak connection with the rhizome they produce thin, broad leaves very easily.
- AAB produces the largest fruit upto 400 mm long.
- AA produces the smallest fruit; 60-100 mm long.

Disease

- Leaf spot, leaf streak or sigatoka disease: The causal organism of the disease is '*Musa cercospora*' which was first identified in Java during 1902 (Asexual stage).

Sexual stage

- Yellow sigatoka – (*Mycosphaerella musical*)
- Black sigatoka – (*Mycosphaerella fijiensis*)

- All AAA clones are susceptible while ABB is resistant.
- Panama wilt or banana wilt: This is a serious soil borne disease caused by *Fusarium oxysporum*. Panama wilt is a serious problem in poorly drained soils.

Curly top, Cabbage top, Bunchy top

- This is the most serious disease of banana in India. It was first reported in Orissa in 1939. It is caused by a virus. The disease was first identified in Fiji in 1891. An aphid (*Pentalonia nigronervosa*) has been identified to be the main agent of transmitting the disease.
- Bacterial wilt or Moka disease – bacterial (*Pseudomonas solanacearum*)

Insect Pest

- Pseudostem weevil (*Cosmopolites sordidus*) and *Odoiporus longicollis* originated in South American to tropical climatic condition, feeds on the ground rhizomes of the banana from April to October and is prevalent in Kerala, Tamil Nadu, Orissa and Mysore.
- Banana aphid – (*Pentalonia nigronervosa*) is a small greyish to green coloured small sucking insect. Vector of the virus causes bunchy top disease.
- Pseudostem borer (*Odoiporus longicollis*)
- Nematodes – *Radopholus similis* and *Pratylenchus coffeae* etc. A total of 71 species of nematodes belonging to 33 generas have been reported to infest banana of which six are pathogenic.
- Appearance of flower bracts instead of leaves; firstly, female bracts followed by male flower (bracts). All bracts bear axillary crescent-shaped meristematic cushions from which the flowers differentiate. Male and female flowers are morphologically indistinguishable until the inflorescence is about 120 mm long.
- **Harvesting** – Banana takes nine months to flower for the first time and at least another three months to mature the fruit. Banana is available through all seasons from September to April.

Yield – Bunch weight and yield (per/g)

Cultivar	Bunch weight	Yield
Dwarf Cavendish	35 (kg)	72 (MT/ha)
Grands Naine	30-40 (kg)	45-65 (MT/ha)
Williams	15-35 (kg)	35-50 (MT/ha)
Chinese Cavendish	34 (kg)	58 (MT/ha)
Robusta	28 (kg)	58 (MT/ha)

- Banana, the largest germplasm collection is maintained at IIHR Bangalore followed by TNAU, Coimbatore.
- AA group introduced clone Pisang Lilin has been used in the breeding programme for imparting disease resistance.
- Krishna Vaghai is a black stemmed sport Hill Banana.
- Moongil is considered to be a mutant of Nerdran.
- ABB group is the most complex of all.
- Muthia syn. Kothia is a widely grown cultivar in Bihar.
- *M. balbisian* produced the highest number of pollen grains i.e. 47142 per another followed by 40119 in *M. acuminate*.
- Dwarfing allele (gene) is dominant in bananas.
- H 95 (Matti × Tongat) has orange flesh which is resistant to sigatoka and nematodes.
- Bunchy top disease, caused by virus, was first identified in Fiji in 1891, and India at Travancore (Kerala) in 1940 from Sri Lanka.
- The disease is also named curly top / cabbage top.
- The dwarf varieties of bananas are very susceptible to bunchy top.
- Infected suckers spread disease to healthy plants.
- The causal agent (virus) is transmitted by the banana aphid, Pentalomia Nigronervasa.
- It is treated with Control (B.T), use of 1% solution of 2,3,5-friphenyl tetrazolium chloride.

Breeding in banana is a difficult exercise due to complexities resulting from parthenocarpy, sterility, polyploidy and vegetative propagation.

Objectives of breeding and improvement

1. Induction of dwarfness.
2. Developing resistance to sigatoka leaf spot.
3. Developing resistance to *Fusarium* wilt.

11

Citrus

Mandarin group	:	Loose skin Orange or Santra
Common name	:	Santra in India
		Batangas mandarin in Philippines Ponkan in China
Botanical name		
Citrus reticulata	:	Common Mandarin
Citrus deliciosa	:	Willow leaf mandarin
Citrus nobilis	:	King of Kunembo (Japan Mandarin)
Citrus unshiu	:	Satsama (Japan)

Citrus reticulata

- A highly polyembryonic species of Chinese origin
- Mandarin occupies 43% area and 42% under citrus spp.
- Seeds of citrus plants do not have dormancy so they should be sown immediately after extraction.
- Blooms three times a year.
  a) February flowering – Ambe bahar
  b) June flowering – Mrig bahar
  c) October flowering – Haste bahar
- Mandarins are highly susceptible to waterlogging.
- Rootstock for HDP – Troyer citrange (1.8 × 1.8 m2)
- Best time for pruning – Last winter or early spring.
- Mandarin, Sweet orange, Acid lime, Grapefruit are highly polyembryonic.
- Pummelo, Tahiti lime, Citron: Monoembryonic.
- Rangpur lime – Most promising rootstock for mandarins and sweet orange.
- Rootstock – Ada Jamir (*Citrus assamensis*) is resistant to greening.
- Citrus fruits have a special kind of fruit skin referred to as 'Leathery rind'.

- Citrus is a micro nutrient loving plant.
- Trifoliate orange – resistant to *Phytophthora* and nematodes.
- Limolin – glycoside is responsible for the bitter taste of citrus fruit Juice.
- Nagpur mandarin was introduced in India in 1894 by Shuji Raja Bhosle.
- Irrigation requirement of mandarin is higher than other citrus spp.
- The state of Andhra Pradesh has the highest production of citrus fruits in India followed by Maharashtra.
- Harvesting period :
    - Feb-March-Central
    - Dec-April -South India
    - Dec-Feb-North India
    - Aug-Oct-Nilgiri Hills
- Storage temperature – 8-10°C
- Storage life – 2-3 months
- Propagation method – T-budding
- Spacing: 6.0 m × 6.0 m
- Planting time – June-July
- The optimum temperature for growth: 22-26°C
- Major states production wise: Madhya Pradesh>Punjab>Maharashtra> Rajasthan>Assam
- Major states productivity wise: Punjab (23.40MT/ha)>Karnataka (21.64MT/ha)> Tamil Nadu (19.31MT/ha)> Madhya Pradesh (17.37 MT/ha)
- Area (India) - 428.31 Thousand ha.
- Production (India) - 5101.21 Thousand MT.
- Productivity (India) - 11.91 MT/ha

  World citrus producers: China>Brazil>India>Mexico>USA
- Satsuma group – (Seedless) Commercial mandarin of Japan.
- King group – Fruits of this group are many-seeded, mature late and have a strong flavour.
- Willow – leaf group – Polyembryonic
- Nagpur Santra – (Ponkan) finest mandarin in the world, grown in Satpuda hills in Maharashtra, Polyembryonic.

- Coorg mandarin – Most important commercial variety in South India.
- Kinnow mandarin – (*Citrus nobilis* × *Citrus deliciosa*). Developed by HB Frost in the USA in 1935. Introduced in India in 1959.
- Khasi mandarin – Locally known as Sikkim or Kamla mandarin, seeds 12-16 polyembryonic.
- Ponkan – It is a popular cultivar of South China and Formosa.
- Clementine – This cultivar originated in Algeria. It is an early maturing monoembryonic cultivar with good quality fruits.
- Beauty – It is highly popular in Australia with the tree showing an alternate bearing tendency.
- Emperor and Fuebelles - Introduced from Australia
- Sutwal – Introduced from Nepal
- **Cultivars:** Nagpur, Coorg, Khasi mandarins of India and Ponkan of China
- ***Citrus deliciosa*** –originated in the Mediterranean region and highly polyembryonic.
- **Cultivars** – Willow leaf mandarin, Kinnow, Wilking of USA and Belinda, an Algerian selection come under the species.
- ***Citrus nobilis*** – a polyembryonic species, native to Indo-China Kunembo of Japan, King of orange of USA. It is a natural tangor.
- ***Citrus unshiu*** – A polyembryonic species of Japanese origin. The famous Satsuma mandarin of Japan and Qwari, Kara, Silver Hill etc.
- Sikkim is the only place for mandarin is packed in wooden boxes.

The aroma of fruits

- Grapefruit – Nootakatone
- Lemon – Citral
- Orange – Valencene
- The acid present in fruits – Citric acid

Orange group

- *Citrus sinensis* – (Osbeck) sweet or common orange, highly polyembryonic species of Chinese origin.
- Sweet orange of introduced in – 1498.
- Harvesting period –

- Dec-Feb (North India).
- Oct-March (South-India)

- Area (India) -184.62 Thousand ha.
- Production (India) -3265.83 Thousand MT.
- Productivity (India) -17.69MT/ha
- Production wise states – Andhra Pradesh>Maharashtra>Telangana>Madhya Pradesh
- Productivity wise states- Andhra Pradesh (24.17 MT/ha)> Madhya Pradesh (17.39 MT/ha)> Karnataka (16.95 MT/ha)> Telangana (14.89 MT/ha)
- Storage temperature – 2-5°C
- Storage life – 2-3 months
- Propagation method – T-budding
- Spacing – 6.0 m × 6.0 m
- Planting time – June – July
- The optimum temperature for growth: 16-20°C
- Inflorescence – Cymose, Vesicle
- Mosambi – Most popular in Maharashtra, Best rootstock for mosambi– Rangpur lime.
- Satgudi – Most popular in Andhra Pradesh, Best rootstock for satgudi– Rough lemon.
- Pineapple – Mid season variety, good for processing.
- Washington Navel – Most popular in California, good for processing.
- Hamlin – This is an early cultivar grown in Punjab, Uttar Pradesh and Haryana.
- Malta Blood Red – Most popular in North India.
- Best rootstock for Malta Blood Red– Karnakhatta, Jatti Katha.
- Shamouti – Seedless variety
- Valencia – Late-season variety
- Jaffa – Midseason variety
- Mudkhed – Bud mutant of Nagpur mandarin
- Preharvest fruit drop is common in citrus plants. It is due to physiological and pathological factors.

- Control measure – Spray of 2, 4-D (20 ppm), most prone var. Mosambi and Blood Red.
- Degreening of citrus fruits is done by calcium carbide
- Best time for pruning – late winter or early spring.
- Sweet orange is susceptible to waterlogging and phytophthora root rot
- Double ring method best for irrigation.
- Deficiency of zinc along with N is the major nutritional problem of sweet orange.
- Pineapple and Valencia – an indicator of greening.
- Cultivars – Mosambi, Malta Blood Red, Sathgudi of India, Valencia, Pineapple, Washington Navel of the USA, Shamouti of Israel, Succari of Egypt, Dobla Fina of Spain.

Sweet orange cultivars have diverse origin e.g.

- Mosambi – Mozambique
- Sathgudhi – India
- Pineapple – Florida
- Washington Naval and Hamlin-Brazil Joffa and Shamouti- Palestine
- Valencia late- China

Acid group (Lime/Lemon)

- Harvesting period- July-September
- Storage temperature – 8-10°C
- Storage life – 6.8 weeks
- Propagation method – Seed
- Spacing- 4.0 × 4.0 m2
- Planting time – June- July
- The optimum temperature for growth – 20-26°C
- Area (India) - 286.20 Thousand ha.
- Production (India) -3148.47 Thousand MT.
- Productivity (India) -11.00 MT/ha
- Production wise states – Gujarat>Andhra Pradesh>Madhya Pradesh> Karnataka >Odisha

- Productivity wise states- Karnataka (23.37 MT/ha)>Andhra Pradesh (16.11 MT/ha)> Madhya Pradesh (15.12 MT/ha)>Telangana (13.27 MT/ha)> Gujrat (13.09 MT/ha)
- Polyembryonic present – Citrus
- India ranks 1st and 17.20% production among major lime and lemon producing countries.
- World producers- India>Mexico> China>Argentina>Brazil>Spain
- Apomixes – Non-recurrent / Nucellar budding
- Polyembryonic citrus – Nucellar
- Pollination (Insects) entomophilous
- Edible part – Juicy placental hairs
- Ethylene production – very low (<0.1) μl C_2H_4/kg/hr.
- Respiration rate – low (5-10) mg of CO_2 release.
- Bearing habit – Mix bearing – bears laterally, flower bud simple.
- Standardized to the tree injection of tetracycline an antibiotic to control the greening virus via control of insect vector (Psylla) population.
- West Indian lime – (Kagzi lime)
- Paramalini and Vikram – This is the clonal selection developed at Marathwada Agriculture University, Parbhani, MH, India.
- PKM-1. This cultivar has been evolved through intensive screening of acid lime clones of Pereyakulam.
- Seedless – This is a late maturing cultivar and now selection is being made in Himachal Pradesh.
- Tahitian lime (*Citrus latifolia*) this is believed to have been introduced from Tahiti to California, Florida and in other growing regions.
- Pramalini – Canker tolerant
- Sai Sarbati – Tolerant to *Tristeza* and canker
- Jai Devi – Pleasant Aroma
- Sweet line – Non-acid juice, resistant to greening, native to India, self-incompatible.
- Rangpur lime – *Citrus limonica*, Native – India, rootstock for Nagpur mandarin, Coorg, Mosambi, Satgudi oranges.

- Kagzi lime is the indicator plant for Tristeza and it is highly susceptible to this disease.
- Citrus canker is the most damaging disease of acid lime.
- Acid-lime is a tropical plant.
- Lemons are divided into 4 grapes:
  - i) Eureka
  - ii) Lisbon
  - iii) Anomalous
  - iv) Sweet lemon

  (i–iv) Origin of South China
- Gajanimma (*Citrus pennivesiculata*) is the most promising rootstock followed by rough lemon for acid lime.
- Classification of citrus was given by Tanka & Swingle (1945)
- Spain is the largest exporter of citrus fruits.
- Ultra dwarf rootstock of citrus – Flying dragon.
- Grapefruit is also known as Forbidden fruit and breakfast fruit.
- Fruit morphology – Hesperidium
- Rate of respiration – Non-climacteric
- Relative salt tolerance – Highly sensitive
- Pollination – Self-pollination (Homogamy)

Lemon

- Eureka, Lisbon, Galgal, Assam lemon
- **Pant lemon-I**: It is a selection from Kagzi Kalan at Pantnagar. It is self-incompatible
- Villafranca – Belongs to Eureka group.
- Nepali oblong, Nepali round

Sweet lime varieties: Mitha Chikma, Mithota

- ***Citrus limon*** - A weakly polyembryonic species of East Asia Origin. Seed cotyledons are white. and in India are not a true lemon.
- Common Name - Lemon, Nimbu, Pahari Kaggi
- Cultivars - Eureka and Lisbon (USA) Femminello and Monachello (Italy) Bernia (Spain)

- Lemon oil is one of the most important citrus oils used for flavouring purposes in soft drinks.
- ***Citrus jambhiri*** - A highly polyembryonic species of Indian origin.
- It is fairly tolerant to many citrus virus diseases, including *Tristeza*.
- Common name - Rough lemon, Khatti, Kada Narangi
- ***Citrus aurantifolia*** - A highly polyembryonic distinctive species of great commercial importance. The origin is belived to be probably India.
- Common name – Key lime, Khata limbu, Kagzi and Sour lime.
- Grown all over India.
- The Kagzi line is the most important commercial cultivar of India. The cultivar is highly susceptible to *Tristeza* virus and bacterial canker disease.

  Citrus limettioides – **A highly polyembryonic species of Indian origin.**
- Common Name – Indian Sweet lime, Limeenhelw, Mitha Nimbu.
- Cultivars: Mitha Nimbu or Sharbati of India belongs to this species.
- It is commercially grown in countries like Egypt and has been employed as rootstock.
- Grown in Central and Northern India.

  Citrus medica – A monoembryonic species of Indian origin.
- Common Name – Citron, Turanj
- Susceptible to frost and found all over India.

  Citrus karna – A moderately polyembryonic species widely used as a rootstock in Northern India having Indian origin.
- Soh-Sarker of Assam belongs to this species.
- Found all over India, useful rootstock for citrus in heavy soils especially for grapefruit.
- Common Name: Id-lemon, Karna orange, Karna, Karna Khata.

Wild, semi-wild species and related genera

Citrus indica **Tanka** – Origin of India

- Common Name – Indian wild orange
- Found wild at high altitudes in Assam.
- Does not survive in the plains.
- Seeds are very big and cover the major portion of fruit.

Citrus latipes – Origin in the eastern part of India.

- Common name – Khasi papeda, found in Assam
- Fruit medium-sized, subglobose, moderately juicy and acidic.
- A cold-tolerant species.

Citrus macroptera – Origin probably in the India-Burma region.

- Common name – Anampapeda, Melanesian Cakoda
- Common plant in Khasi Hills, Assam.
- Fruits are medium-sized, subglobose, very juicy and highly acidic in taste.
- Two forms: Satkara and Tithkara.

Citrus ichangensis

- Common Name – Ichang Papeda
- Petiolar wing well-developed, fruits are inedible. A cold-hardy species.

Citrus assamensis – Origin of India

- Common name – Adajamir
- A distinctive species having crushed leaf aroma similar to ginger or eucalyptus smell.
- Found semi-wild in Assam. It is similar to the Gajanimma or Baduapulli of South India and should be treated as a botanical variety of *Citrus pennivesiculata*.

Citrus aurantium - (Tree in India), Highly polyembryonic species of Indian origin. The species was primarily used as one of the principal rootstock and flowers are used for perfumery purpose.

- Since the species is intolerant to certain citrus viruses like *Tristeza*, in more recent years, it has been abandoned as rootstock.

Pummelo - grapefruit group

- *Citrus grandis* – pummelo or shaddock: A monoembryonic species of South East Asia origin (largest fruit in citrus) self-incompatible.
- Common name - Chakotra
- Cultivars: Kao Pan of Thailand, Buntan of Formosa
- *Citrus paradise* - Grapefruit or pomelo: A polyembryonic species of southern China and West India, origin.
- Common Name -Grapefruit
- Cultivars - Foster, Ruby, Marsh, Duncan (Seedless, Thompson).

- Cross protection, a few commercial nurserymen now use the technique for the immunization of nursery plants of acid lime with the milder strain of the *Tristeza* virus.
- Use of trifoliate orange (*Poncirus trifoliate*) for the breeding of rootstock for resistance to Phytophthora causing collar rot and nematodes.
- Producing countries of the world in grapefruit & Pummeto.
- Blood Red, Joffa, Valencia and Pineapple commercial grow in Punjab.
- Hong Kong Kumquat (*Fortunella hindsii*) is a tetraploid and Tahiti lime (*Citrus latifolia*) is a seedless triploid.
- Mandarins, Sweet orange, Acid lime, Grapefruit are highly polyembryonic.
- Pummelo, Citron and Tahiti lime are monoembryonic.
- Wood apple, *Feronia limonia*, is a highly dwarfing rootstock for citrus suitable for high density planting precious but is not recommended due to short life of grafted plants.

Related genera

- ***Poncirus trifoliata*** – Origin of China
- Common name – Trifoliate orange
- Plants deciduous with trifoliate leaves.
- Grown in India as a hedge plant.
- Used as a rootstock for citrus in Japan
- *Fortunella margarita*
- *Fortunella japonica*
- *Fortunella crassiflora*
- *Fortunella hindsii*
- The fruits are used for the preparation of candy and the plants for ornamental purpose.
- The small evergreen shrub can withstand cold weather.

  Hybrids and cultivars: Intergeneric and intrageneric hybrids

Intergeneric

A. Hybrids of *Poncirus*

   i) Citrange – (*Poncirus trifoliata* × *Citrus sinensis*)

      The hybrids show the intergeneric character of the parent cultivar – Troyer, Carrizo, Mortex, Etonia Rush, Coleman etc.

ii) Citrangequat - *Poncirus* × *Citrus* × *Fortunella*

iii) Citrangedin - *Citrange* × *Citrus mitis* (Calamondin)

iv) Citrangor - Developed by back cross (*Citrange* × *Citrus sinensis*)

v) Cicitrange - Another back cross (*Citrange* × *Poncirus trifoliate*)

vi) Citrumelo - *Poncirus trifoliate* × *Citrus paradise*

vii) Citrandarin - *Poncirus trifoliate* × *Citrus reticulata*

viii) Citramon - *Poncirus trifoliate* × *Citrus limon*

ix) Citradia- *Poncirus trifoliate* × *Citrus aurantium*

x) Citrumguat - Very difficult hybrid to breed

(*Poncirus trifoliate* × *Fortunella japonica* × *Fortunella margarita*)

B. Hybrids of Fortunella

i) Procimequat – *F. japonica* × *C. aurantifolia* cv. Mexican × *F. hindisii*

ii) Limequat - *C. aurantifolia* × *F. japonica*

iii) Orangequat - *C. reticulate* cv. Satsuma × *F. japonica* × *F. margarita* cv. Meiwa

Intrageneric

i) Fangor - *C. reticulata* × *C. sinensis*

Cultivar - Temple, Clementine, Monreal, Umatila are some important cultivars, mostly monoembryonic

iii) Fangelo - *C. reticulate* × *C. paradise*

Cultivar - Orlado, Sampson, Minneola, Seminole.

iv) Lemonime - *C. limon* × *C. aurantifolia*, Cultivar- Parrine

v) Lemomage - *C. limon* × *C. sinensis*

vi) Lemondrin - *C. limon* × C. *reticulata*

Grapefruit varieties

- Duncan – This cultivar is considered to be the oldest grapefruit and grown in Florida.
- Marsh seedless – It has been replaced by Duncan in Florida and India.
- Thompson Seedless – Seedless
- Red Blush – It is synonymous with Ruby Red Seedless. Red Marsh
- Saharanpur Special- it has been developed from Saharanpur.

Rootstock

- Rangpur lime – Susceptible to foot rot burrowing nematode and tolerant to *Tristeza*, suitable for heavy and deep soils.
- Karna Khatta – Suitable for heavy soils used in U.P. incompatible with Kagzi lime and Malta Blood Red.
- Rough lemon – Commonly used in Punjab, susceptible to blight and tolerant to *Tristeza*.
- Trifoliate orange – Cold hardy, resistant to phytophthora, Tristeza and nematode, dwarfing and salt-resistant suitable for shallow and high rainfall areas, incompatible with lime, lemon and mosambi.
- Flying dragon – Most dwarfing 'Fortunella'
- *Severinia buxifolia* – Salt resistant
- Soh Sarkar – dwarfing for Kinnow
- *Feronia limonia*- Dwarfing for but not commercial
- *Citrus unshiu* – Freeze tolerant
- Cleopatra mandarin – Cold hardy, tolerant to tristeza and nematodes.
- Citranges –tolerant to gummosis and tristeza.
- Sour orange – Cold hardy, suitable for sandy loam soil, susceptible to tristeza.
- Dwarfing rootstocks- Thomasville, Citrangequat, Feronia and *Severinia buxifolia*.
- Calamondin is an orangutan developed through a cross between (*C. reticulate* × *F. crassifolia*)
- Temple is a natural tangor and originated in Jamaica.
- Pummelo (*Citrus grandis*) is considered to be the ancestor of grapefruit (*Citrus paradise*)
- Grapefruit originated in Barbados (West India). It is a forbidden fruit hence named as *C. paradise*
- Citrumelo is a hybrid between Trifoliate orange × Grapefruit is an ultra-resistant rootstock which is immune to Tristeza tolerant to exocytic, phytophthora, gummosis, excess salinity and cold.
- Soil – Wide range of soils, pH to between 5.5 and 7.5.
- Usually, a period of 3 to 4 weeks is required for germination in a secondary nursery bed.
- The optimum temperature for germination ranges from 25-30°C.

- Carrizo citrange is used as a rootstock for Lisbon and Eureka lemon to wet infection and incompatibility.
- Similarly, sprays of 2,4-D at 60ppm applied in June or July reduce fruit drop during summer or early fall.
- High temperature also increases the accumulation of sugars in tunnel house fruits during stage 1 growth.

Disease

- Gummosis – *Phytophthora* spp.
- Bacterial Canker (*Xanthomonas campestris* pv. *citri*)
    - Transmitted by leaf miner.
    - Resistant variety Tenali
- Tristeza (Virus) Acid lime indicator plant Transmitted by aphids.
- Greening (MLO's), Indicator plant – Sweet orange Transmitted by citrus psylla
- Exocortis (Viroids) Rangpur lime and Citron-indicator.
- Xyloporosis – Bud wood transmission.

Pest

- Psylla (*Diaphorina citri*), Vector of greening disease

  Control- Parathion – 0.025% / Malathion – 0.05%
- Leaf miner – (*Phyllocnistis citrella*) Vector of citrus canker

  Control- Rogor – 0.1% / Metasystox – 0.03%
- Aphids – (*Toxoptera aurantii*) Vector of *Tristeza* disease.

  Control- Malathion – 0.05% / Phosphamidon – 0.025%
- Lemon butterfly (*Papilio demoleus*) Control by bagging and Malathion – 0.05%

Physiological Disorders

- Granulation – High temperature & RH during ripening.

Spray of lime

- Leaf mottling / frenching – Zn deficiency
- Exanthema / Ammoniation dieback / - Cu deficiency
- Yellow leaf of citrus – Mo deficiency

12

Coconut

Common Name	-	Kalpavriksha
Botanical Name	-	*Cocos nucifera* Linn
Family	-	Arecaceae / Plames
Chromosome No.	-	32, X=16
Climatic adaptability	-	Tropical
Origin	-	South-east Asia
Fruit morphology	-	Drupe
Dichogamy	-	Protandry
Pollinating agents	-	Wind and Insect
Area (India)	-	2096.70 Thousand Ha
Production (India)	-	16412.57Thousand MT
Productivity (India)	-	7.83 MT/ha
Top producer in India	-	Kerala> Karnataka> Tamil Nadu> Andhra Pradesh> West Bengal Top Productivity in India - Tripura (16.43 Mt/Ha)> Andhra Pradesh (9.68 Mt/Ha)> Telangana (9.40 Mt/Ha)> West Bengal (8.61 Mt/Ha)> Kerala (8.35 Mt/Ha)
Top Producing in World	-	El Salvador> Guyana> Tonga> Nepal> Peru
Propagation	-	Seed
Spacing (m2)	-	7.5 m × 7.5 m
Planting time	-	May-June and Oct-Nov
Inflorescence	-	Spadix, (1.2-1.8m long)
Edible part	-	Endosperm
Bearing habit	-	Axillary on current season
Yield	-	80-100 nuts/plant/year

Bearing habit - Bears fruit all along the length of old wood in the leaf axil.

Soil - Sandy soil and pH

- Coconut Research Station (CRS)-1916.
- Central Coconut Research Station, Kayangulam, Kerala (CCRS)- 1947.
- Generic name (cocus) is derived from the Spanish word 'coco' which means 'monkey face'.
- Coconut industry employs 10 million people.
- Monolaurin content of coconut oil has anti-HIV property.
- Dwarf coconut is self-pollinated while and tall coconut is cross-pollinated in nature.
- Fully mature nut has 30-40% coir.
- Kerala has the highest share (45%) in coconut production in India: followed by Tamil Nadu (22%) and Karnataka (12%).
- The productivity of (7608 nuts/ha) Indian coconut is the best in the world.
- India ranks no.1 in terms of coconut productivity in the world.
- In India, 9-12 months old seedlings are generally transplanted
- Coconut is derived from the Spanish word 'coco meaning 'monkey face'.
- Tall palms: Var. typical (Nar), most commonly cultivated in all the coconut growing areas of the world, the height of 25-30m. Long pre-bearing age of 6-10 years, cross-pollinated.
- The fruit is generally medium to large in size and nuts mature within 12 months.
- The copra content is usually over 150g/nut and the oil percentage varies from 66 to 70%.
- Dwarf palms propogate through self-pollination: Var. nana (Griff).
- They are quick to come to bearing (3-4 years).
- They have thin trunks without a swollen base or 'bole' and fully developed fronds rarely exceeding 4m.
- In India, three important dwarf types are found ; (1) Chowghat Green Dwarf and (2) caught orange dwarf mainly described by the colour of their nuts and petiole, from Kerla and (3) Gangabondan a gran dwarf from A.P.
- The copora content in dwarfs ranges from 90g to 120g/nut and oil is about 65%.
- At present (PCRI, Kasaragod is maintaining the world's largest assemblage of coconut germplasm).

- CPCRI is the World coconut Germplasm Centre is situated at Sipighat in Andamans.
- Indian National average coconut yield – 36 nuts / palm / year.
- Mother palm selection high yielding 80 nuts and 80 kg per palm per year.
- Chandrakalpa yielding 97 nuts/palm/year was released by CPCRI-1985 (Syn. Laksha Dweep Ordinary).
- Banawali Green Round (Syn. Benaulin) from the Goa region has been selected and released in 1987 by Kankan Krishi Vidyapeeth, Dapoli (M.H) for cultivars in Kankan coast under the name Partap yield 151 nuts/palm/ year).
- Hybridization work was started in 1932 by the crossing of west coast tall X Chowghat green dwarf.
- Chandra Sankara (CODXWLT) in 1985 for Kerala.
- Root wilt, the most serious problem in coconut MLOS, is controlled by the application of 1% bords pest.
- The shell resembles two eyes and nose monkeys.
- The dried kernel is commonly called 'Copra'.
- Highest productivity A.P.
- Each root on an average measures 6m to the top and 1.5m from ground level. They do not have root hairs.
- Each root lives up to 20 years and 5-10 year old (red coloured) roots are more active.
- No cambium tissue is found in coconut trees.
- The grown is an adult palm consists of 25-35.
- Stomata are confined to the lower surface of the leaf. As 12 leaves are shed every year, thus age.
- It exhibits monoecious flowering which commences at 6 years of age and is borne regularly.
- Genetically, the dwarfs are considered autogamous (self-pollinated) and the tall ones, allogamous (cross-pollinated)
- Generally, about 25-40% of the female flower reaches maturity.
- National yield of coconut 36 nut / palm / year.

o Malayam dwarf yellow	
o Malayam dwarf green	These are the introduction of Malaysia.
o Malayam dwarf orange	

- TXD hybrid comes to bearing in about 4-5 years.
- Seed nuts will commence germination within 8-10 weeks after sowing.
- Potash is the predominant nutrient requirement of the palm.
- Coconut husks are rich in potash.
- A new method of fertilizer application is known as pocket mooring. Tea Board – 1954 Kolkata.
    - Rubber Board – 1947 Kottayam
    - Cottee Board – 1942 Bangalore
    - Spices Board – 1987 Cochin
- Wet meat or Kernel: The kernel or endosperm.
- Dry (Desiccated coconut) at a temp of 77° to 82°C for 40-45 mins.
- Coconut water is the liquid endosperm of the tinder coconut; it takes 5-7 months for making a refreshing drink.
- Toddy: Toddy is a sugar-containing juice obtained by tapping the unopened spadix. On an average, about 15-18 liters of toddy is extracted in one month containing 5-8% alcohol
- Milling copra: Copra with moisture content of less than 6% and free from disease and pests is used for the production of oil is referred to as milling copra.
- Coconut oil: Coconut ousting by expellers leaves about 7-8% oil in the cake refining to lower free fatty acid (FFA). Out of total oil produced, 50% is used as having and skin oil 35%.
- Edible purposes and the remaining 15% is used for industrial purposes.
- Coconut fiber (Coir): It is an important product obtained from the fruit husk. The fully matured (12-month-old) nuts yield about 35-40% coir. 11 month old nuts have a still higher coir content ranging from 50-55% while that of 9-10 month old nuts is as high as 70%.
    - 50% hair & skin oil 35% edible oil
    - 15% of industrial purposes
- Age – oil%
    - 2 months – 35-40%
    - 5 months – 50-55%
    - 10 months – 70%
- Coconut contributes to six per cent of the edible oil output in the country.
- The coconut industry survives mainly on coconut oil.

- Utilized for production of milling copra is around 36% which forms only 12% of the mould total production of copra.
- Kerala accounts for 75% of the total products of coir.
- Sri Lanka is one of the major contributors exporters of desiccated coconut worldwide.
- In India, the per capita availability is low i.e. 11 nuts/year, 215 in the Philippines, 60 in India and 151 in Sri Lanka.
- Cross-pollinated and heterozygous in natural.
- "Double century" AICRIP workshop was held during 1995.

Two forms of copra

1. Edible copra
   i) Ball copra
   ii) Cup copra
2. Milling copra

- In Kerala, 60-65% of the total coconut produce in converted into Milling Copra.
- Mesocarp – Husk – used for coir making.
- Coconut husk
  i) Coir (70%)
  ii) Fiber (30%)
- Endocarp is used for making toys, buttons etc. Secondary growth in coconut stem as well as root is absent.
- Coconut is a heliotropic plant (loves sun shine).
- Kurumba – An immature coconut containing a refreshing clear liquid.
- In drip irrigation, 30-40 litters water/day is optimum for west coast condition.
- Coconut ripens in 12-13 months from the opening of the inflorescence.
- 55% of coconut produce is consumed raw.
- Laccadive ordinary is suitable for making ball copra and oil extraction.
- Laccadive Micro is also suitable for making ball copra.
- Kangayam (T.N.) is considered as the major market centre for copra coconut oil in the country.
- The first hybrid between Tall and Dwarf coconut was released in the year 1932.
- The fully mature nut has 30-40% coir.

Tall varieties

- West coast tall, Laccadive ordinary, East-west tall, Andaman ordinary, Sacrament, Pratap, Benalin tall, Laguna

Dwarf varieties

- Chowghat green dwarf, Chowghat orange dwarf, Gangobondan, Gudanjali, Loco Nino, Mangipod, Nuleka-Cross pollinated
- Hot spot areas have been surveyed and disease-free palms identified and a resistance breeding programme is being initiated.
- Crop mechanization includes the development of various types of copra driers, electronic copra moisture meter, coconut de husher bunch supporter and tree climbing device.
- The coconut palm is monoecious and bears spikelets (30-35) each having about 250-300 male flowers at the top and one of the few female flowers at the base.
- The dwarf cultivars commonly used as a male parent are dwarf orange, dwarf green, Gngabondam and Malayan dwarf yellow.
- Dwarf orange and Gangabondam area better male parents for economic hybrid production.
- Horizontal planting of nut is better than vertical planting.
- The palm is monoecious (Female and Male flowers in the same inflorescence)
- The inflorescence of coconut is spadix. Spicata palm has unbranched inflorescence.
- Dwarf palms are autogamous while tall types are allogamous.
- NCD stands for natural cross dwarf hybrids. They bear early and yield high.
- Multitier cropping system is an important concept in coconut orchards.
- The tapping of coconut is done from unopened inflorescence for 'Toddy'.
- Coconut oil constitutes 7.0 per cent of the annual vegetable oil requirement.

Composition and uses

- The wet meat or kernel – The endosperm of the mature coconut is known as meat or kernel.

  Moisture – 45%, Protein 4%, Fat – 37%, Carbohydrate – 10%, Minerals – 4%.

Water coconut

- The water inside the tender or green coconut is known as Coconut Water. It acts as an antiseptic to the urinary tract and increases blood circulation in the kidneys.

  Water – 95.5%, Carbohydrate – 4%,
  Protein – 0.1%, Minerals – 0.4%
  Fat – 0.1%, Calcium – 0.02%.
  pH – 4.8 to 5.3

- Use – Vinegar, Nala-de-coco beverage may be prepared from the coconut water.
- The Lakshadweep micro variety is very suitable for the manufacture of ball copra.
- Cup copra is prepared by cutting the fully matured dehusked nuts into halves and drying under the sun.

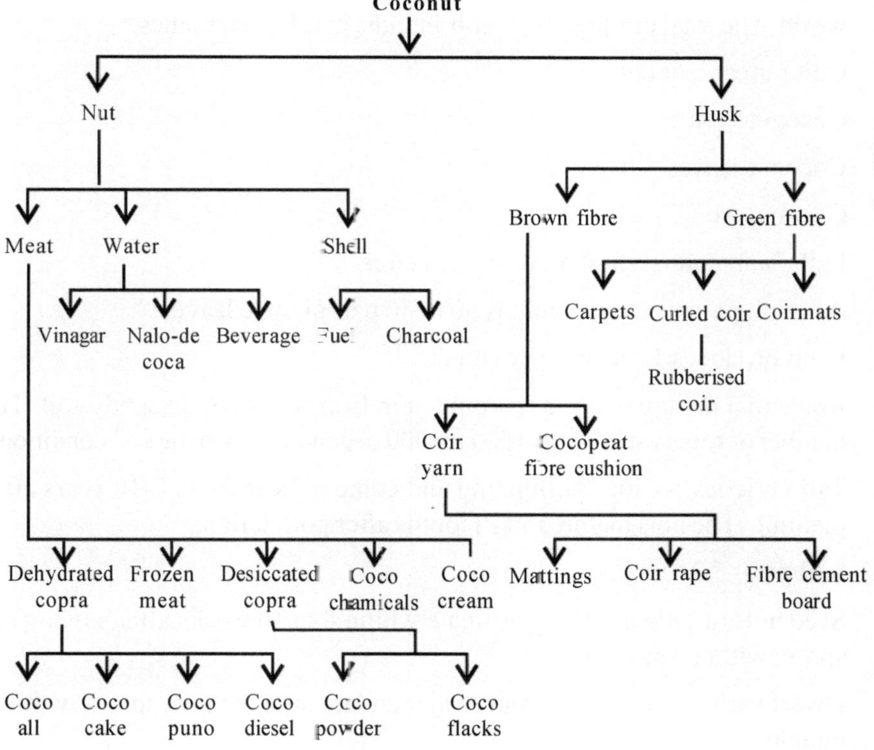

- Coconut milk and cream
- Coconut flour

- Coconut oil – Coconut oil is extracted from dried copra by crushing. The oil varies from 58 to 65%.
- Coconut cake – The cake or poonac left after extraction of oil from copra forms about 34-42%. It contains 8 to 17% residual oil.
- Toddy and toddy products – Toddy or sugar-containing juice is obtained from the coconut palm by tapping or puncturing the unopened spadix.
- The yield of toddy per palm per day id about 1.5 litres. A healthy palm can yield 300-400 litres of toddy per year.
- "Triacle" is a concentrated golden coloured syrup which is produced by boiling toddy.
- It contains 5-8% alcohol. Distillation of the fermented toddy yields a strong liquor known as "Arrack".

Coconut husk

- Coir or coconut fiber – India is the premier coir producing country in the world. The total production of coir in India is 2,11,200 tonnes.
- Coir pith or coir dust
- Coconut shell
- Coconut leaves
- Coconut wood
- Fall plant attains a height of 15-30 meters.
- An adult palm usually contains 30 to 40 pair pinnate leaves.
- Palm produces 12 leaves per annum.
- Roots of a mature tree range from 5m in firm soil to 7m in sandy soil. The number of roots varies from 1500 to8000 depending upon the soil conditions.
- Tall varieties – Cross-pollinating and come to bear fruits 7-10 years after planting. The nuts mature in 12 months after pollination.
- Soil pH – 5.2 to 8.0.
- Seed nuts of tall varieties germinate within 8 to 10 weeks after sowing nut sprout within 5 months.
- Dwarf variety seed nuts sprout earlier and most of the nuts sprout within 3 months.
- Seedlings transplanting 9 to 12-month-old seedlings.
- Application of BHC 10% on the husk prevents termite attack.

- Observed that regular inter cultivation increases 32.9 per palm per year.
- Floral Biology – Pollen discharge or anthesis continues for about 18 to 20 days in an inflorescence.
- The pollen output of a healthy male tree about 1, 11,000 to 2, 21,000 grains. The pollen remains viable for about 2 to 9 days.
- The female flowers commence opening within 18th to 21st day.
- The fertilized flower takes 11 to 12 months to develop into a mature fruit, popularly known as nut.
- A fully matured fruit has a composition of about:
- Husk – 35%, Shell – 12%
- Kernel and Meat – 28 Water – 25%
- D breeding and Varietal Improvement
- The manifestation of heterosis or hybrid vigour in coconut was first reported in India during 1932.
- Pests – Rhinoceros beetle – (*Oryctes rhinoceros*)
- Fan like appearance of leaves is caused by larval parasites – *Bracon bremic* real control

  – 0.1% BHC.
- Red palm weevil – (*Rhynchophorus ferrugineus*) Carbaryl (Sevin)– 1%
- Eriophyid mite

Disease

- Root wilt – (MLO's)
- Transmitted by lace bug.
- Basal end rot (fungus) *Ganoderma lucidum*
- Thanjavur or Ganoderma wilt
  - First noticed in Thanjavur district in 1952-1955.
  - Crown choking – physiological disorder due to boron deficiency 50g/ palm – Borax.
- One of the most beautiful and useful trees in the world. It is grown in more than 80 tropical countries.
- The root system of monocots produces numerous thick roots from the base of the stem continuously almost throughout its life.

- The fruit is large reaching a length of 20 to 30cm. The outer layers of the pericarp are thick and fibrous.
- Pollination –Dwarf palms are considered autogamous (direct self-pollinating) while the tall ones are allogamous (cross-pollinating)
- All India coordinated coconut and arecanut improvement project was approved in 1970 and started functioning in 1972.

13

Pineapple

Common Name	-	Ananas
Botanical Name	-	*Ananas comosus*
Family	-	Bromeliaceae
Chromosome No.	-	50
Climatic adaptability	-	Tropical
Fruit morphology	-	Sorosis
Rate of respiration	-	Non-climacteric
Acid tolerance	-	Medium tolerant
Self-incompatibility	-	Gametophytic
Pollination	-	Cross-pollination
Center of origin	-	Brazil
Acid present	-	Citric acid
Largest producing State in India	-	West Bengal
Year of Introduction in India	-	1448
Harvesting period	-	May-August
Area (India)	-	102.96 Thousand Ha
Production (India)	-	1705.76 Thousand MT
Productivity (India)	-	16.57 MT/ha
Top producer in India	-	West Bengal> Assam> Karnataka> Meghalaya> Manipur
Top Productivity in India	-	Karnataka (62.42 MT/ha)> Telangana (50.00 MT/ha)> Maharashtra (31.88 MT/ha)> Tamil Nadu (31.12 MT/ha)> West Bengal (30.26 MT/ha)
Top Producing in World	-	Costa Rica> Brazil> Philippines> China India

Top Productivity in World - Indonesia (132.15 MT/ha)

Export from India - Nepal> Qatar> Saudi Arabia>UAE> Maldives

Storage temperature - 7.5-12°C, RH 70%

Storage life - 7 weeks

Propagation methods - Suckers, slips and crown

Spacing (m2) - 22.5 × 60 × 5 cm3

Planting time - April – June Optimum temperature for growth - 22.32°C

Inflorescence - Spike

Pollination - Birds (Ornithophilous)

Edible part - Bracts / Perianth

Ethylene production - Low – 0.1-1.0μLGH4/kg/hr Fruit

Bearing habit - Old season growth (Terminal)

Enzyme - Bromelin

Moisture Content - 68.71-80.10%

Sugar - 10-12%

Protein - 0.69-1.89%

Fat - 0.6-0.57%

Carbohydrate - 15.8-28.95%

Fiber - 0.95-2.5%

- Fresh seeds germinate in 20-30 days giving as high as 90% germinate.
- An NPK mixture supplying 375g N + 202 P2O5+375kg o/tree gave the highest mean-field of 33 fruits/tree and 14.88 kg fruit/tree.
- Flowering planting after four years.
- Flower bud differentiation of at 40 leaves stage.
- Flowering time March-April after 7 weeks of fruit development.
- Mealybug attacks shoots and in between the fruit segments – 0.5% phosphamidon or dichlorvos.
- A virus transmitted by lace bug or mealybug causes pineapple wilt.
- Ripon normally can be stored at 5°C for six weeks.
- Suckers weighing 500-750g and slips weighing 300-400g are better for yield and quality.

- Pineapple exhibits vegetative parthenocarpy and self-incompatibility.
- This is non-climacteric fruit.
- Sunscald discord is due to exposure of fruit to sun rays. Multiple crown/ fasciations is another important disorder probably due to genetics/nutritional factors.
- It contains an enzyme known as Bromelin.
- The fruit which matures in winter is acidic.
- Pineapple is a monocotyledonous, monocarpic, CAM and herbaceous fruit plant.
- It has introduced in India in 1548.
- The tree bears commercial fruit at the age of 6-7year.
- Fruit weighs between 800g to 2kg.
- Cultivar – Pink's Mammoth, African Pride Bradley, Page, Gefner, Island Gem.
- Queen – Mauritius cultivar
- Favourable soil conditions range from heavy soils to almost sandy texture. Sandy, marginal and wastelands may be utilized for pineapple cultivation.
- Annonas is mostly a subtropical fruit preferring warm climate.
- Flowers are layered at 20/15°C day/night temp.
- Pineapples are propagated using of suckers and slips.
- Optimum temperature for growth is 22-32°C.
- Optimum planting time is April-June
- Optimum spacing is 22.5 cm × 60 cm × 75 cm with 63400 plants per/ha.
- Fresh seeds germinate in 20-30 days giving as high as 90% germinate.
- NPK mixture comprising 375g N + 202 P2O5+375kg o/tree gives the highest mean-field of 33 fruits/tree and 14.88 kg fruit/tree.
- Flowering start after planting of four years.
- Flower bud differentiation at 40 leaves stage.
- Flowering period is March-April after 7 weeks of fruit development.
- Mealybug attacks shoots and in between the fruit segments – 0.5% phosphamidon or dichlorvos.
- Pineapple wilt is caused by a virus and transmitted by lace bug or mealybug.
- Ripon can be stored at 5°C for six weeks.

- Suckers weighing 500-750g and slips weighing 300-400g are good for yield and quality.
- Pineapple exhibits vegetative parthenocarpy and self-incompatibility.
- Pineapple is a non-climacteric fruit.
- Sunscald discord is caused due to exposure of fruit to sun rays. Multiple crown/fasciations in another important disorder probably caused due to genetics/nutritional factors.
- It contains an enzyme – Bromelin.
- The fruit which matures in winter is acidic.
- Pineapple is a CAM, monocarpic, herbaceous plant.
- Ethereal (Ethephon) is used for inducing flowering in pineapple.
- Average sugar content is 10-12%.
- Acid content is 0.6-0.8%.
- Plant density of 63,400 plants /ha (25 × 60 × 75cm3) is ideal for subtropical and mildly humid conditions.
- Earthing up is an essential operation in pineapple cultivation.
- Pineapple does not contain starch.
- NAA and NAA based compounds *viz*. Planofix and Celemone @10-20ppm induce flowering in pineapple but they are less effective. So ethereal is used.
- Multiple crown disorder is found in the cayenne group. (Kew)
- Sugarloaf – Sweetest and best-flavoured fruit.
- Propagation – Slips (300-400g); Suckers (500-750g)
- Aluminium sulphate – Best N2 fertilizer for pineapple.

Varieties

- Kew – Leading commercial variety valued for canning (NER) particularly for canning late variety.
- Giant Kew – Northeastern region (canning)
- Charlotte Rothschild- This pineapple variety has characteristics similar to Kew and Queen. This variety is partly under cultivation in Kerala and Goa
- Queen – Early variety grown on hills; best dessert cultivar (NER).
- Mauritius – Midseason variety of green group turns red after ripening mainly grown in Kerala.
- Jaldhup & Lakhat – Both under green group.

- Jaldhup has characteristic alcoholic flavour.
- Cayenne – Tropical variety cultivated commercially in the Philippines. (2n =3x=75)
- Other popular varieties: H1, H3, H7, H8
- Grown in Singapore, Malaysia for the canning industry.
- Induction of uniform flowering by the chemical combination of 25ppm Ethephon + 2% urea and 0.04% sodium carbonate applied at 30-40 leaf stage NAA at 200-300 ppm enhances fruit weight.
- Nair is a herbaceous perennial.
- The fruit is syncarpous (multiple) fruits.
- Pineapple exhibits vegetative parthenocarpy and self-incompatibility.
- Self-incompatibility is a major breeding hindrance in pineapple.
- Flatbed planting is suited for western plains while furrow planting is practiced in hills.
- Ethephon and other ethylene releasing compounds are more commonly used for floral induction in pineapple.
- NAA (10-15ppm), B - hydroxyethyl hydrazine ethylene chlorohydrins, ethrel etc., induce uniform flavouring (or flowering) in pineapple.
- Mealybug pseudococcus (Dysmicoccus) bee wipes is a serious pest that transmits a virus which causes wilt.
- Heart rot is caused by *Phytophthora parasitica.* The infection is severe in wet and dry regions and is more predominant in alkaline soils.
- Sunscald – a disorder in pineapple is caused due to exposure of fruits to sun rays.
- Multiple crown/fasciation is another important disorder caused probably due to genetical/nutritional factors.
- Fruits are used both as fresh and processed products.
- A commercial source of protease is “Bromelain”.
- The enzyme is active in hydrolyzing protides or peptides over a pH range of 5.5 to 8.5.
- Excellent source of vitamin C and a good source of vitamin A and
    - Water – 77-91%
    - Total acidity – 3.8-7.0g%
    - Sugar – 9.7-12.1%
    - Protein – 0.36-0.50g%
    - Vitamin B – 10-14 µg/100g

- 1500 A.D onwards, the plant was proliferated widely and rapidly throughout the old world.
- The mature plant is 1.0 m to 1.2 m high and 1.3 to 1.5m in diameter.
- The stem (meristem) produces 70 to 80 leaves unless it is prematurely induced to bloom. The time between planting and formation of the inflorescence varies between 6-16 months.
- Flowers open early in the morning and close by evening. The lower flowers on the inflorescence open first then over about 3 weeks flowering proceeds gradually up the inflorescence axis.
- The pineapple fruits are usually parthenocarpic and seedless.
- The fruit tapers towards the top where it is mounted by a rosette of short, stiff, spirally arranged leaves called the "Crown".
- Ananas is typically diploid (2n=2x=50), tetraploid (2n=2x=100)

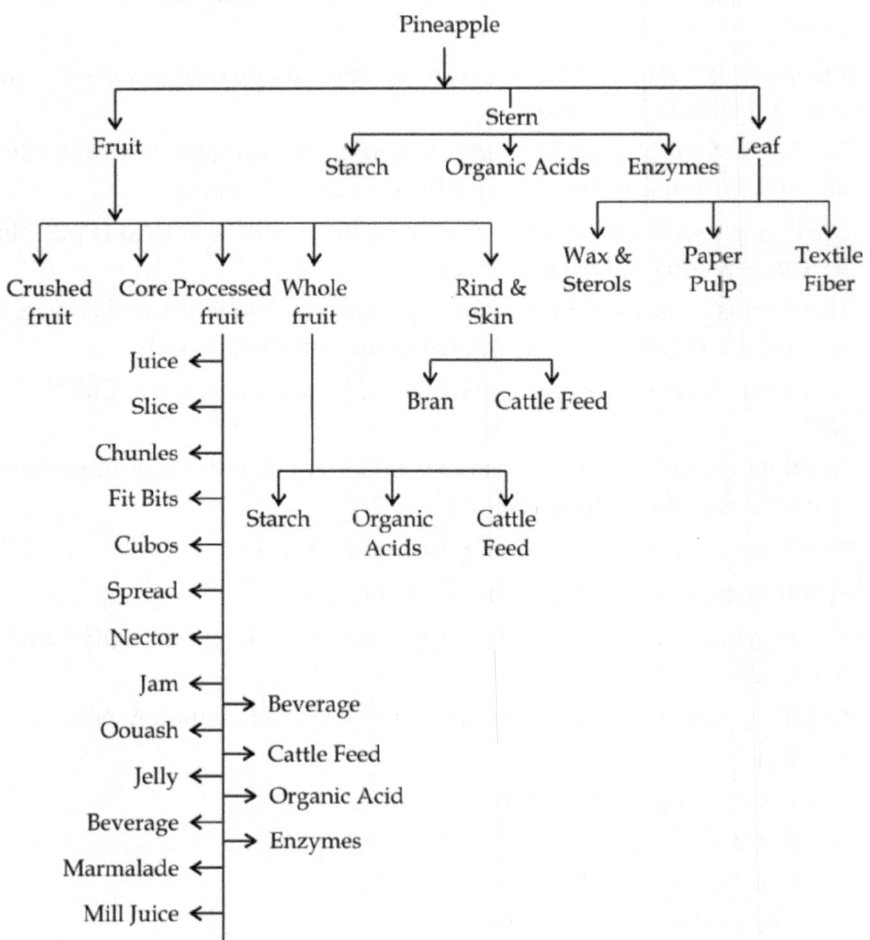

Five groups of cultivars

- Spanish groups – Shape of fruit is globose, with large deep-set eyes. Fruit weight from 0.9-1.8kg. The rind is deep reddish-orange coloured, flesh pale-yellow to white, a spicy-acid taste and fibrous texture. It is resistant to Mealybug.
- Queen Group: An F1 plant takes four years to produce fruit and in the case of an interspecific cross four backcross generations will probably be necessary to produce a plant suitable for commercial use.
- One cultivar of pineapple "Cabazon" is a natural triploid.
- Triploid and tetraploid pineapple remains self-incompatible and is hence seedless.
- In India, pineapple is grown commercially in Assam, W.B. Bihar, Coastal Andhra Pradesh, Kerala, Karnataka, Orissa, Goa, Tripura, Meghalaya and Tamil Nadu.
- Optimum rainfall ranges from 1000 m -1500mm.
- The pineapple root system is very sensitive to waterlogging.
- Slips produce fruit after 18 to 20 months.
- Suckers produce fruit/plant in 15-18m.
- Crown produces plant in 22-24 months.
- Suckers slips are usually planted in 10-15cm deep holes.
- A mixture of 16g N, 4g P and 16K per plant produces the highest yield.
- Recommended at 10:6:10 NPK mixture @ 300kg/ha at 3 to 6 months.
- Bromacil is very effective as a pre-emergence herbicide against nutgrass.
- Diuron @ 2kg/has and Bromacil @ 3kg/ha reduced both dicot and monocot weed population.
- A period of 6 months before setting of the natural fruit appears to be satisfactory for the treatment.
- The cut end of the fruit stalk may be dipped in a solution of 10% Benzoic acid in Alcohol to check any of fungus.
- Fruits are graded into four classes:

  A. 1500-1800g

  B. 1100-1500g

  C. 900-1100g

  D. 700-900g

- Mealybug – (*Dysmicoccus brevipes*) the symptoms first appear on the roots and gradually spread over the remaining plant.
- Treat the basal portion of sucker or slips with 0.04% solution of methyl.
- Wilt – (Virus) Transmitted by mealy bug.
- Yellow spot – (Virus) transmitted by thrips from hosts as such.

Physiological disorders

- Sunscald – Exposed surface gets damaged due to exposure to direct sun rays. Can be prevented by covering the plant with dry straw or banana leaves during peak summer months.

  Fasciation and multiple crowns –

  Multiple crowns may occur due to genetical factors as well as due to soil and environmental reasons, sometimes the fruits get flattened and fascicled.
- Blackheart – Also known as endogenous brown spot, or internal browning. Caused due to exposure to high temperature – 40°C, Control – Treatment with 150-300ppm GA3.

  Ripening – Perola with ethereal at 500-2000 ppm results in the development of uniform colour within 8 days.
- Pineapple Research Station of the Bidhan Chand's Krishi Viswavidyalaya at Mohit Nagar in Jalpaiguri district of West Bengal.
- Application of NAA (200-300) in the form of Planofix within 2-3 months after fruit set increases fruit set by 15-20 per cent.
- Urea (2%) and CaCo3 (0.04%) induces more than 90% flowering after 50 days of application.
- Flower stage is attained within12 months after planting and formation of at least 40 leaves.
- Sandy, loamy and laterite soil on hilltops are best for pineapple.
- 'Kenyan Kathal' a hardy plant producing seeded fruits with non-edible sour and scant pulp can be included in the improvement programmes.
- Pineapple is a xerophytic, succulent, herbaceous, perennial, mono-cotyledonous plant with dense rosette leaves.
- The leaves yield 2-3% of strong silk fibers; (38-90 cm long) which are used for making a fine fabric called pina cloth.

14

Papaya

Common Name - Melon tree

Botanical Name - *Carica papaya* L.

Family - Caricaceae

Chromosome No. - 18, 36, X=9

Climatic adoptability - Tropical

Fruit morphology - Berry

Rate of Respiration - Climacteric

Photoperiodic Response - Day-neutral plant

Pollination - Self-pollination

Center of origin - Tropical America (Mexico)

Growth curve - Double sigmoid curve

Area - 138.40 Thousand Ha.

Maximum production - 5988.80 Thousand MT

Maximum productivity - 43.30 MT/ha

Top Producers (India) - Andhra Pradesh>Gujrat> Karnataka>Madhya Pradesh> Maharashtra

Top Productivity (India) - Andhra Pradesh (93.72MT/ha) Tamil Nadu (92.83MT/ha) Karnataka (67.84MT/ha) Telangana (66.93MT/ha) Gujrat (61.86MT/ha)

Top producing (World) - India> Brazil> Mexico> Indonesia> Dominical Republic>Nigeria

Top Productivity (World) - Dominical Republic (431.60 MT/ha), Indonesia (90.61 MT/ha), Mexico (56.59 MT/ha), Brazil (46.91 MT/ha), India (43.30 MT/ha)

Introduction in India - 1611

Storage temperature - 9-10°C

Propagation by - Seed

Spacing (m^2)	-	1.8 × 1.8, 3 × 3
Planting time	-	Sep-Oct
Optimum temperature for growth	-	24-27°C
Largest melon producing country	-	Brazil
Inflorescence	-	Cymose (Solitary)
Pollination	-	Wind (Anemophilous)
Edible part	-	Mesocarp
Ethylene production	-	High – (10-100) μLC_2H_4/Kg/hr.
Bearing habit	-	Axillary (Current session growth) (Laterally on Shoot)
Sex ratios	-	B& 1: @& 20 (but in Pusa Delicious 50:50)
Papain contain	-	72.2% protein
Seed rate	-	250-300 g/ha (Gynodioecious), 400-500 g/ha (Dioecious)
India's rank in terms of production	-	1st and share 44.40%
Yellow pigment in papaya	-	Caricaxanthin

- The enzyme present in dried latex of papaya (papain) is pepsin.
- Frost is the most limiting factor for papaya cultivation in North India.
- Papaya is a polygamous plant. –
- 20 seeds of papaya weigh one gram
- Carpine obtained from papaya is used as a diuretic and heart stimulant.
- 10% male plants are planted wherein dioecious varieties are calculated.
- Papaya plants are extremely susceptible to waterlogging.
- Papaya is thermosensitive crop.

  Carica candoamarcensis is scientific name for Mountain papaya.
- Papaya has the highest production after banana.
- Sunrise Solo type papaya produces no male plants.
- Seeds are enclosed in gelatinous sarcotesta.
- Irrigation is done using ring methods.
- Pusa Nanha can have 1.25 m × 1.25 m (6000 plants/ha) planting density

Gynodioecious varieties

- Pusa Delicious - (E & N part of India) Resist nematodes.
- Pusa Majesty - One of the highest papain yielder (Eastern & Northern part of India)
- CO-3 (T.N) - (IIHR) Coorg Honeydew - Selection from honeydew, first hermaphrodite of Madhu Bindu
- Sunrise Solo - Pink flesh, pear-shaped, weighing 350-500g and exhibits excellent quality.
- Taiwan - Blood red in colour
- *C. papaya* is a polygamous plant and has three basic sex forms size staminate, pistillate flower and Hermaphrodite (monoecious).
- The sex reversing male plant can produce all 8 types of flowers during the year. The hermaphrodite flower (plants) can produce 6 types except for staminate and teratological staminate flowers.
- The use of PGR has been found to change sex in papaya to increase femaleness.
- Spraying with GA3 (50ppm) has been reported to increase femaleness and phosphonate D or SADH-250ppm.

Genes

M_1^{RR} - Dominant factor for homozygous sex reverse male.

M_1R^2 - Dominant factor for heterogynous sex reversing male.

M^{rr} - Dominant factor pure maleness.

M_2 - Dominant factor for pure maleness.

M_2 - Recessive factor for femaleness.

Genetic constitute

$M_1\ ^{RR}m$ – Sex reversing homozygous male plant.

$M_1\ ^{Rr}m$ – Sex reversing heteronyms male plant.

$M_1\ ^{rr}m$ – Pure male plant

M_2m – Hermaphrodite mm – female plant.

- Length and ramification of the inflorescences are secondary sex characters.
- Crossing among the basic sex forms in papaya and ratio of segregating progenies.

- A dominant gene for suppressing femaleness (SUF) is present only in the pure male. As a result, this plant never produces any fruit. The suppressor ger (SUF) is absent in sex reversing males. Thus, all such plants are capable of bearing fruits.

Crosses	Female (mm)	Hermaphrodite (M_2m)	Male (M_1m)	Non-viable (Lethal)
mm x M_1m	50^4_{1x4}mm	-	50M_1m	
mm x M_2m	50_1 ^{mm}	50_1M_2m		
M_2m x M m	25_2mm	50_1Mm	25M_2M_2	
M_1m x M m	25_1mm	-	20Mm	25_1MM
M_2m x M m	25_1mm	25_1Mm	25_1M_2m	$25_1M_1M_2$
M_1m x M_2m	25_1mm	25M_2m	25M_1m	25M_1M_2

Diecious varieties

- Pusa Giant - Suitable for making tutti-frutti icecream and candies, used in the canning industry and has good wind resistance.
- Pusa Dwarf - Eastern and Northern part.
- Pusa Nanha - Extremely dwarf, suitable for HDP (Pot garden)
- CO-1 - (TN) Dwarf and resistant to nematodes.
- CO-2 - (TN) for papain.
- CO-6 - (TN) Selection from Pusa Majesty for papain.
- CO-5 - (TN) Cultivated mainly for papain extraction (Papain yield 1500-1600 kg/ha)
- Pink flesh sweet: TSS - 12-14°Brix
- Other varieties: Pant - C1, Harts Gold, Sunny Bank, Betty

Hybrid varieties

- CO-3 (TN) - CO-2 × Sunrise Sola
- CO-4 (TN) - CO-2 × Washington
- CO-7 (TN) - CO-3 × Pusa Delicious × Coorg Honeydew.
- Seed can be stored at 10°C and storing in airtight bottles to maintain viability over 9 months.
- Seeds treatment with thiourea 100-200 ppm or GA_3 200 ppm improves seed germination.
- Pusa Majesty – Tolerant to viral disease and root-knot nematodes.

- *Carica cauliflora* - is a source of resistance for papaya ringspot virus (PRSV)
- Spacing closer than 1.6 × 1.6 is detrimental to papain production.
- Pusa Nanha – is a mutant developed through 'gamma irradiation' of seeds of a local variety.
- Honeydew – Suited for North India heavy bearing, less no. of seeds with good favour and test.
- Drip irrigation in papaya saves around 50-60% of water. Its water requirement is estimated to be 1800-1900 mm and irrigation at 60-80% available soil moisture depletion is optimum.
- *Carica quercifolia* is the hardiest of all the species.
- Papaya (*Carica papaya*) also called papaw or pawpaw, is a quick-growing typically single-stemmed, short-lived large perennial herb.
- Young leaves are eaten in Java as a vegetable.
- The genus Carica is native to Mexico.
- Papaya being a quick-growing, continuously fruiting, evergreen plant requires good fertile soil. A well-drained sandy loam soil, rich in plant food, with pH - 5.5-7.0 is ideal for papaya cultivation.
- Plants are highly sensitive to flooding and waterlogging may cause the death of plants in 3 to 4 days.
- The waterlogging symptoms are defoliation of old leaves and chlorosis of the remaining leaves.
- The favourable temperature for Papaya is between 21-33°C.
- The incidence of malformation is high on fruits from flowers initiated during low winter temperature (<16°C) and the cv. Sunrise Solo is highly susceptible to this disorder.
- Shows optimum growth at 60% humidity.
- Mixture of bisexual and female plants in the ratio of 2:1 is recommended for the production of papaya.
- Highest Seed germination capacity (40%) after 3 years has been observed in Coorg Honeydew.
- Tissue culture – Rapid multiplication of true-to-type, disease-free desirable genotypes of papaya clones by micropropagation has been attempted successfully.
- Prevention of 'wet feet' is crucial during papaya irrigation.

- Potassium deficiency symptoms include yellowing and necrosis of lower leaves.
- Application of paraquat at 0.5-1.0 lb/acre (0.-23-0.45kg and karmex (diuron) at (2.5-5.0) lb/acre (1.13-2.26) kg.
- Flowering – Three sex forms are commonly found in papaya i.e. distillate, hermaphrodite and staminate.
- Floral Biology – Peak anthesis was found between 5am and 6 am. in all species except in the distillate flowers. Female and male flowers develop within 32 and 42 days respectively of bud initiation.
- Sex expression – Linnaeus (1753) classified papaya as a dicecious species.
- Spraying with GA3 and SADH is reported to induce femaleness while application of MH, SADH and Phosfon-D reduce the plant height, accelerate flowering and fruiting at lower node, increase fruit set and yield. Application of CCC and TIBA induces advance flowering at lower nodes and increases fruit set.
- The fruit quality, yield and size, papain yield and the photolytic activity of papain can be increased by treatment with GA_3.
- Depending upon the quality of Cultivars, the papaya fruit normally takes 145 to 165 day to attain edible ripe stage from the date of flowering.
- Fruits are harvested between 5:00 am and 7:00 am.
- The first crop of fruits is harvested within 12-14 months after transplanting.
- Storage of the fruit is possible for up to four weeks at 8°C for the fruit of minimum maturity at harvest.
- Papain – The immature papaya fruit contains milky latex.

  The dried latex called 'Papain' has international markets in the UK and the USA. Papain is used for tenderizing meat, clearing beer, manufacturing of cosmetics like snow and face creams and dental paste. Papain is used in detecting stomach and intestinal cancers and in correcting diphtheria.

  Unripe but fully-grown fruits yield maximum papain. 2½-month-old fruits yield the highest quality of papain. The papain yield is more than tapped at 90 days after fruit set. Papain yield is higher from July to January. Yields may be as high as 6g of papain per fruit. The cv. 'CO_2' recorded the highest papain yield of 686.29g per plant in six months. The application of ethephon (100-200ppm) IBA (25 or 50ppm) and (GA3 – 100-200ppm) also increases latex production.

- Coorg Honey Dew performs well in Karnataka but is poor yielder in North India.
- Papaya originated as a cross between two species of the genus (Carica) native to Mexico. The first record of papaya was made by the Dutch Traveler, Linsochotex in 1576, who observed that the plant had been brought to Malacca from the West Indies.
- The productivity of the fruit in India is 30t/ha contributing 35% to the world production.
- Papaya + tobacco intercropping is common in North Bihar.
- Crops such as chillies, tomato and brinjal should not be grown with papaya since these can as hosts to viruses.
- No intercrop should be taken once the flowering and fruiting starts.
- Cool and wet period produces more papain.
- After collecting, the latex should be dried in the sun or electric oven at 40°C in case sum drying is not feasible.
- Potassium Metabisulphite (0.05%) is good for preservation of papain
- The annual yield of papain varies from 200-300gm per tree with a maximum of 450gm. The yield per hectare works out to 250-375kg per year.
- Papaya came to India via Malacca in the 16th century.

Disease

- Damping-off – Fungus (*Pythium aphanider* Matum, *Rhizoctonia* spp. *Fusarium* spp.

Control

- Ringspot (Virus) is transmitted through aphids and causes mottling of leaves also known as mycoplasma.
- Leaf curl (Virus) is transmitted through white fly. Also known as mycoplasma.
- Collar (*Pythium aphanidermatum*), food or stem and root rot (*Phytophthora palmivora)*
- Control: Bordeaux mixture – 1%, Copper oxychloride – 2g/litre of water.
- Powdery mildew (*Qidium caricae*), Control: Bavistin and Bayleton – 0.1%, Dithane M – 45 (0.2%)
- Bunchy top – *Cladosporium oxysporum*) the pathogen invades through the injury caused by thrips.

Pest

- Aphids – (*Aphis gossipi*) Transmitted through mosaic virus
- Nematodes- (*Meloidogyne incognita*) Root-knot
- In the main field, 25g of Furadon may be applied per plant.
- Red spider mite – Control – Lime sulphur.

Disorders

- Skin freckles – Fruit freckle index and freckle diameter are the lowest during the late summer-autumn part of the year. Wrapping young fruit in white paper bags significantly reduces incidence of freckle.
- Dieback – The plants display a brown discolouration of the vascular tissue.
- It fruits in about a year, gives the highest production of fruit per unit area and generates income next to the banana.
- Under ideal conditions, apples and grapes give more income but not higher yields.
- Seed germinates in three to four weeks.
- The cutting root well after treatment with three to five per cent IBA
- Four to five leaves should be transplanted in June.
- Transplanting in December gives fruits on the plant when they are much shorter.
- The larger leaves of the seedlings are cut off at the time of transplanting.
- The plants flower in about 10 months in North India and 5 months in South India.
- 240g N, 500g P, 500 K, and 25 kg FYM.
- In North India, harvesting starts December onwards. In the Central and South India, it is done in February.

15

Pear

Common name	-	Pear
Botanical name	-	*Pyrus communis*
Chromosome No.	-	34
Climatic adaptability	-	Temperate / Deciduous Fruit morphology - Pome
Rate of Respiration	-	Climacteric
Relative salt tolerance	-	Highly sensitive Self incompatibility - Gametophytic
Origin	-	Europe (Western China)
Harvesting Period	-	May-June
Yield	-	30-35 MT/ha
Storage temp	-	-1.0°C
Storage life	-	4-5 months
Propagation	-	Tongue grafting / T-budding Largest producer in the world - Italy
Inflorescence	-	(With leafy shoots) Corymbose
Pollination	-	Insects (Entomophilous)
Edible part	-	Fleshy thalamus
Respiration rate	-	Medium (10-20mg) of CO_2
Bearing habit	-	Axillary – old season growth
Fruit bud	-	Mixed bud
Flowering habit	-	Seasonal flowering fruits
Growth pattern	-	Single sigmoid growth curve
Colour pigments	-	Orange-yellow – flavonoids Genome size China - 5/2mbp (*Pyrus bretschneideri*) 2012 Flower colour - White

- More tolerant to wet soils but less tolerant to drought than apple.
- Browning of pears is due to polygalacturonase activity.

- Major acid – Malic Acid
- Chilling – China Pear (Low chilling) – 150 hrs.
- European pear – 1200-1500hrs.
- Most commonly used clonal rootstock – Quince A cutting.
- Quince is scientifically – Cydonia oblonga origin – South-Eastern Europe and Asia minor.
- Quince is a monotypic genus
- Vigorous rootstock – *Pyrus pashia*
- Graft incompatibility is overcome by double grafting with old home or hardy varieties.
- In India, the open center system is common
- Training system modified central leader system
- Double working and double grafting in pear
- Most of the pear varieties are self-sterile (due to gametophytic SI)
- Most of the pear cultivars grown in hills are partially self-fruitful
- Introduction from Europe: Bartlett, Anjou, Kieffer
- Low chilling Varieties – Kieffer, Leconte, Patharnakh, Gola, Punjab Nectar
- High chilling Varieties – Anjor, Bartlett, conference, Flemish beauty
- Soft fleshed – Red blush, Punjab Gold, Punjab Nectar.
- In North India – Nashpati, Pathamakh
- Temperate region – Bartlett (1500 hrs)
- In Tamil Nadu – Kieffer (Kodaikanal)
- Free from gritty cells – Flemish Beauty (Polli-variety)
- Spontaneous mutation (Bud sports) – Starkrimson, Clapp's Favourite (Red coloured pear)
- The red colour and flavoured variety – Starkrimson D. Red Arjan from Arjun.
- Interspecific hybrid – Leconte and Kieffer (*P. communis* × *P. serotina*)
- Bartlett or Williams or William Bartlett is the most popular variety all over the world.
- Kieffer: Well adopted widely grown in India
- PAU – Punjab Gold, Punjab Nectar, Punjab Soft

Intergeneric sterile hybrid

- Mule – Troth early peach × wild goose plum
- Kamdesa – Peach × send cherry
- Pyronia – Pear × Quince
- Resistant to fire blight – *P. Calleryana*
- European pear is highly susceptible to fire blight of pear than an oriental pear.
- Pear decline is caused by MLOS which is transmitted by pear psylla.
- Storage disorder – Core breakdown and scald.
- Free from storage disorder – Anjou.
- Boron deficiency – Corky tissue, Calyx-end rot and blossom blast.
- Black-end and cork spot is due to deficiency of calcium.
- Core breaks down on brown heart is due to abnormal cool season.
- Hard end of pear is due to unfavourable water conditions.
- The pink end is due to abnormal cool-season preceding harvest.
- Willo leaf pearl (*P. salicifolia*) is frost, drought and salt resistant rootstock.
- Mehal / Kainth – *P. pashia* important
- Shaira – *P. serotina* used as rootstock
- Spring frosts are detrimental to pear production and temperature at – 3.3°C or below kills the open blossoms.
- Quince – A is most commercially used rootstock production trees of 50-60% of standard size.
- For active growth of pear, soil depth should be 180cm.
- Bartlett is famous for dry pear slices.
- Bartlett – Baggughosha (Interspecific hybrid) Willian's Bartlett (England)
- Flemish beauty – Self-fertile, good pollinizer.
- Kiffer – Cross between French pear × oriental pear self-fruitful, processing suitability, can be grown in South India.
- Patharanakh (Sand Pear) – Self-pollinated.
- Varieties free from Git cell – Flemish beauty, Magness
- High N2 is not suitable: because the incidence of pear psylla and fire blight is more
- Premature ripening: Pink colouration near blossom end. It is due to Night temp. <7.1°C, day temp <21°C

• Chinese centre	*P. pryrifolia*
(North, Central China	*P. ussuriensis*
Japan, Korea)	*P. betulifolia*
	P. calleryana.
• **Central Asiatic centre**	
Western Tian-Shan,	*P. communis*
Uzbekistan,	*P. pashia*
North-West India and Afghanistan	*P. salicifolia*
• **Near Eastern Centre**	
Asia minor	*P. communis*
Caucasus mountains	*P. syriaca*
	P. laucasica
• **European group:**	
	P.communis
	P. nivalis
	P. cordata
	P. caucasica
• **Asian group:**	
	P. calleryana
	P. betulifolia
	P. dimorbhophylla
	P. kochnei
• **North Africa Group:**	
	P. longipes
	P. memorensis
	P. gharbiana
• **East Asian Group:**	
	P. pyrifolia
	P. ussuriensis
	P. hodoensis

- Both European and Asian pears were domesticated by hybridization and selection from local wild species in prehistoric times.
- Chinese and Japanese migrants took Asian pear to the USA as seed and scion material.

- Fruits some times during 1970 and planted them in Kullu, Shimla in (H.P.) and Kumaon hills in UK.
- H.P. has a maximum area of 7382 ha with a production of 17381 MT.
- The plant is deciduous tree or shrub, leaves senate, senate rarely lobed.
- Fruit – a globose or pyriform, pome with persistent or deciduous calyx.
- Pear can be tolerated as low a - 26°C temp and high 45°C during growth period.
- A pH range of 6.0 to 7.5 is desirable because Fe deficiency appears in alkaline soils.

Early season varieties

Early China, Laxton's Superb, Fertility Thumb pear, Borne of Amanlis.

Midseason

Bartlett, Red Bartlett, Max Red Bartlett Starkrimson, Flemish Beauty, Clapp's Favourite, Dr Jules Guyot, Pathenakh, Leconte.

Late season

- Conference, Doyenne Du-cons, Kashmir Pear, Beurre Hardy, Wincor of Winkfield, winter Nelis.
- Many varieties of pear were introduced from England, France. Italy and Japan.
- Clonal rootstocks are Quince A (Vigours) Quince B (Intermediate) and Quince (dwarf) and BA29 OHXF-230 (semidwarf) Oregon 211 dwarfing.
- Seeds are stratified for 30-40 days in moist sand at low temp (4-5°C) during Dec-Jan.
- Cuttings are treated with 100ppm / IBA for 24 hours.

Grafting in Feb and budding April-May.

Name of country	Training system	Planting distance (m)
Japan	Perola	7.5 × 7.5 or 9 × 9
New Zealand	Tatra Trellis	5 × 4
	Centre leaders	7.5 × 7.5
India (Hill)	Modified centre leader	5 × 5
	HDP (Quince-C)	3 × 3
Indian (Plains)	Centre leaders	8 × 8
	Modified centre leaders	6 × 6

- Pear bear fruits mostly on spurs or some time on 2 years old wood-spurs continues to bear for 6 years. The limbs with spurs over 6-8 years old need to be removed in a phased manner.
- NPK ratio of 70:35:70g/year age of the tree and which is stabilized after 10 years of age (700:350:700g)/tree FYM at the rate of 10kg/year and 100kg for 10 years and above.
- Nitrogen will be applied in the month of Feb-Mar in two split closes. Half of N shell be applied 2-3 weeks before bud break and second half-dose one month after flowering.
- Paclobutrazol (PP333) @ 500 to 1000ppm restrict the vegetative growth and increase fruit set in pear CV. Flemish Beauty.
- GA3 at 10-20ppm applied 10-14 days after full bloom increased fruit set, fruit retention but it reduces flower bud differentiation in the following year.
- NAA 5-10ppm control fruit drop in pear.
- The most critical period of water requirement in pear is April to June months and peak requirement is after fruit set.
- Fruit weight may increase up to 20% in delayed picking, however, it reduces the storage life.
- In Washington (USA) harvest maturity – 110-115 days
  - Bartlett – 130-135 days
  - Bosc- 145-150 days
- Generally, Bartlett pear is harvested at a 19-pound pressure.
- Harvesting should be done in 2-3 picking at 3-4 days intervals.

Grading and packing of pear:

Grade	Equatorial diameter (mm)	Size of box (inner in cm)	No of layer
Extra large	<75	45.3 × 30.5 × 30.5	4
Large	70-75	45.3 ×20.5 × 20.5	3
Medium	65-70	-do-	3
Small	60-55	-do-	3
Extra Small	55-60	-do-	-
Culled	<55	45.3 × 30.5 × 30.5	Loose

- Pear is packed in boxes either in offset or in diagonal styles.
- European Pears: Bartlett, Clapp's Favourite, Anjor, Conference, Winter, Flamish Beauty, Anne-du-comice, Max Red Bartlett, Red Bartlett, Starkrimson, Laxton's Superb.
- Asian Pears: Shinseiki, Chojuno, Kosui Nijjiseiki, Kikisu

16

Sapota

Common Name	-	Sapodilla plum, Bully
Botanical Name	-	*Achras zapota / Manilkara achras*
Family	-	Sapotaceae
Climatic adaptability	-	Tropical
Chromosome No.	-	2n – 26.
Fruit morphology	-	Berry
Rate of respiration	-	Climacteric
Dichogamy	-	Protoandry
Pollination	-	Cross-Pollination
Origin	-	Mexico
Growth Curve	-	Double Sigmoid curve
Area (India)	-	97.29 Thousand Ha
Production (India)	-	1175.89 Thousand MT
Productivity (India)	-	12.09 MT/ha
Top producer in India	-	Gujarat > Karnataka> Tamil Nadu > Maharashtra> Andhra Pradesh
Top Productivity in India	-	Mizoram (50.00 MT/Ha)> Tamil Nadu (29.50 MT/Ha)> Madhya Pradesh (17.21 MT/Ha)> Andhra Pradesh (13.03 MT/Ha)> Karnataka (12.46 MT/Ha)
Export from India	-	UAE> Bahrain > Oman > Qatar > Canada
Year of Introduction in India	-	1898 in Maharashtra, Vill. (Gholwad).
Propagation	-	Inarching

Spacing (m2)	-	10.0 m × 10.0 m
Planting time	-	July- August
Optimum temperature for growth	-	11-34°C
Post Harvest Losses	-	9.73%
Inflorescence	-	Solitary (Lymose)
Pollination	-	Wind (anemophilous)
Edible part	-	Mesocarp
Ethylene production	-	Very high >100µLC2H4/kg/ha
Bearing habit	-	Bears fruit all along the length of old wood in the leaf axil.
Largest producer in the world	-	India
Harvesting period in India	-	January-February and May-June
Storage temperature	-	20°C with 5–10% CO_2
Storage life	-	35 days
TSS	-	18%

- Delicious Fruit
- Sapota fruit is a good source of sugar which ranges between 12 and 14 per cent.

Mixture	-	73.7%
Sugar	-	12-14%
Carbohydrate	-	21.4%
Protein	-	0.7%
Fat	-	1.1%
Calcium	-	28 mg

Dried Sapota

Iron	-	27 mg
Ascorbic acid	-	1.39%
Ascorbic acid	-	6 mg

Cultivars

- Kalipatti, Chaatri, Dhola Diwani, Long Bhuni or Bhunipatti, Jingar, Vanjet, Pata Kirthabharthi Dwarapudi, Cricket Ball, Oval, Vavi Valosa, Bangalora Calcutta Round, Jonnavalosa I, II, round Baramari Pot Sapota, Gavarayya, Thagonampadi, Ayyangan (Rose scented)
- Flowers of Sapota and protogynous Pollination is mediated through the wind.
- The most ideal soils are deep alluvium, sandy loamy red late rites and medium black soils.
- Sapota is a tropical fruit crop optimum temperature ranges between 11°C and 34°C.
- Sapota is grafting of inarching.
- Rootstock – Sapota seedlings, Rayan or Khirnior pala Adam's apple, Mahua, Star apple.
- Seed germinates by the 4th week, budding and grafting is done on seedling about 1cm thick.
- Optimum Planting distance 10 m × 10 m

1-3 years	- NPK	- 50:20:75g	FYM	- 40 kg/year
4-6 years	- NPK	- 100:40:150g		
7-10 years	- NPK	- 200:80:300g		
11 years and more	- NPK	- 400:260:450g		

- Flowering appears in March, April and October; Maximum flowering occurs in June during early morning hours i.e. 5 am to 7 am, Max – 8-11 am.
- 22per cent of fruits set Sapota is cross-pollinated.
- Leaf spot by a fungus (*Phaeophleospora indica*) was first reported from Dhawan India by Chinnappa (1968).
- Co1 is a hybrid between Cricket Ball and oval.
- Co1, Co2, Cricket Ball, Kallipatti are resistant to leaf spot of Sapota.
- Calcutta Round is most susceptible to leaf spot.
- Chiku moth (*Nephopteryx eugraphella*) causes 40-60% field dosses.
- Fruit fly (*Dacus correctus*) and fruit borer (*Anarsia achrasella*) are serious pests of sapota.
- Central leader system of training is the most common method of training for Sapota.
  - CO1 - Cricket Ball × Oval
  - PKM2 - Guthi × Kirtibharti

- ° PKM3 - Kallipatti × Cricket Ball
- ° DSH1 - Kallipatti × Cricket Ball
- ° DSH2 - Kallipatti × Cricket Ball

- Appropriate time for harvesting sapota fruits is after 245 days of fruit set.
- Harvesting time: Jan-Feb and May-June on the west coast of Maharashtra.
- Yield – 3 years old plant yields about 100 fruit @ 15-20 T/ha.
- Storage of fruit between 12°C and 14°C increases the storage time to about 5 weeks.
- Ripe fruit can be kept at 2°C -3°C and 85-90 per cent R.H. for 6 weeks while the firm fruits should can be stored at 3-5°C for 8 weeks.
- The fruit is a good source of digestible sugar (12-18%)
- Rayan rootstock of sapota is the best in respect of plant vigour productivity and legality.
- For uniform and rapid ripening, Ethephon (1000ppm) can be utilized at 20-25°C.
- Softwood grafting using rayan as the rootstock is the best method of propagation with 93% success rate.
- Central leader system of training is the most common method of training Sapota.
- Fruits are dipped in GA_3 300ppm + Bavistin 1000ppm solution at the pre-packing stage to extend shelf-life and avoid storage loss.
- Dhola Diwani – This variety was found in Maharashtra and harvested in summer.
- Ayyangar – This is a rose-scented variety and bears large-sized fruit.
- Kirti Barti – Popular in A.P., thick skin, good transport value (egg.-shaped) and very sweet to taste.
- Cricket Ball – Famous in A.P. (Resistant to leaf spot)
- Kalipatti – Popular in MH. (Resistant to leaf spot) harvest in winter.
- Murabba – Popular in MH.
- CO-2 Clonal selection from Baramsi (Resistant to leaf spot).
- PKH-1 Clonal selection from Guthi-Dwarf variety.
- Calcutta special round is most susceptible to leaf spot.
- Pilipath – Suitable for high-density planting.
- CO-1 – Cricket × Oval resistant to leaf spot.

- PKM-2 – Guthi × Kirtibharti
- PKM-3 – Kalipatti × Cricket ball – Dwarf.
- CO-3 – (Cricket ball × Vavivalsa) HDP.
- Rootstock – (*Manilkara hexandra*) commercially used for most of the sapota cultivars (Khirni).
- It can grow up to a height of 1000 meters, but it does not do very well above 500 meters.
- It thrives in areas having 100-150cm rainfall.
- The mortality after transplanting air-layered successfully can be up to 50% and some varieties do not layer successfully.
- The planting is done in May before the rains start.
- PKM1 – Columnar tree shape.
- The unripe fruits and bark yield a milky white later which solidifies on exposure to air and this forms the basis for pickle making while ripe fruits are sweet-smelling and delicious.
- Sugar contents range between 12% to 14%. The pulp is also made into sherds and halves.

Four types of varieties

i) Tree with erect growing habit

ii) Tree with drooping habit

iii) Tree with spreading habit

iv) Tree with spreading habit but fruits inferior

Sapota is grown in deep alluvium, sandy loams, red laterites and medium black soil with pH range – 6-8.

Rootstocks

1. Sapota seedlings (*Achras Sapota*)
2. Rayan or Khirni or pale (*Manilkara hexandra*)
3. Adam's apple (*Manilkara kauki)*
4. Mahua (*Madhuca latifolia and Mimusops kauki*)
5. Mee tree (*Bassia longifolia*)
6. Star apple (*Chrysophyllum cainito*)
7. Miracle fruit (Sideroxylon dulfolin)

   Sapota can tolerate drought conditions to some extent.

Fertilizer

Age of the plant	N	P	K
1-3 years	50	20	75
4-6 years	100	40	150
7-10 years	200	80	300
11 year and more	400	160	450

Yarm yard manure – 40kg/per/ year

- Flowering –Sapota starts bearing small crops from the second or third year of planting but economical yield can be obtained from seventh year onwards. Flowering takes place during March, April and October while maximum flowering in June.
- The peak of anthesis is reached at 4 am. Average pollen size – 16.25µ.
- Sapota has 22% natural fruit set while the maximum fruit drop occurs immediately after fruit set.
- NAA results in a better fruit set than that effected by GA3.
- The seed number per fruit ranges from 1 to 9 and the round fruit has higher number of seeds.
- Sapota harvesting after fruit set – 245 days.
- Field –
    - 3-year-old tree – 100 fruits/year
    - 5-year-old tree – 250 fruits/year
    - 7-year-old tree – 700 fruits/year
    - 8-year-old tree – 800 fruits/year
    - 10-year-old tree – 1000 fruits/year
    - 11-year-old tree – 1500 fruits/year
    - 15-year-old tree – 2000 fruits/year
    - 30-year-old tree – 2500-3000 fruits/year
- Fruit ripens in about 5 days after harvest at room temperature. They ripen well at 12-14°C.
- The optimum storage temperature is 20°C. The storage life can be increased by removing ethylene and adding 5-10 per cent CO2 storage atmosphere.
- Temperature between 2°C to3°C and 85-90% RH are optimum for 6 week storage while temperature range of 3-5°C and 85-90% RH are good for 8 weeks storage.

- For long distant transport (road or rail), the fruit should be treated with 75ppm GA3 for 3 minutes.
- Controlled cross-pollination resulted in the highest fruit set of 78%.

Disease

- Leaf spot – This disease is caused by the fungus; *Phaeophleospora indica* and was first reported from Dharwar by Chinnappa-1968 in India. 0.2% - Dithane Z-78, 0.5% - Blitox
- Sooty mould – This is a common disease reported India by and is caused by fungus of *Capnodium spp*. It is caused due to excretion by scale insects and mealybugs. Treatment: 40g Zenib in 18 litres of water or 100g starch solution 10 litres water.

Pests

- Stem borer (*Indarbela tetraonis*) has been reported from Tamil Nadu and Karnataka.
- Flower bud eating by caterpillar
- Fruit borer-(*Virachola isocrates*)

Disorder

- Cock's comb.

17

Litchi

Botanical Name - *Litchi chinensis*

Family - Sapindaceae

Origin - China

Chromosome No. - 30, X=14,16,15,17

Climatic adaptability - Subtropical

Fruit morphology - Nut

Rate of respiration - Non-climacteric

Relative acid tolerance - Medium tolerant

Relative salt tolerance - Less

Photoperiodic responses - Long day plant

Type of pollination - Cross-pollination

Harvesting period - May-June

Area (India) - 92.3 Thousand Ha

Production (India) - 686.4 Thousand MT

Productivity (India) - 7.4 MT/ha

Export from India - Country-wise Nepal>UAE>Saudi Arabia> Kuwait> Norway

Storage temperature - 5.7°C

Storage life - 35 days

Propagation methods - Air layering

Spacing (m^2) - 8.0 m × 8.0 m (156 plant/ha)

Planting time - June-July

Optimum temperature for growth - 30°C

The largest producer in the world - China

Inflorescence	-	Compound raceme
Pollination	-	Insects (entomophilous), honeybee and white flies
Edible part	-	Fleshy aril
Respiration rate	-	Very low
Bearing habit	-	Terminal – old season growth
Fruit bud	-	Simple
Fruit	-	Single seeded nut

- Air layering is called "Marcottage" in China and "Gootee" in India.
- IBA (2-10g/liter of water) is most effective in root promotion in air layering of litchi.
- Flowers are petalless.
- Subtropical evergreen fruit tree; prefers a moist climate.
- Moist summer and cool winter favour litchi production.
- Red pigment – Anthocyanin.
- The pulp is an outgrowth of the seed in litchi.
- Roots – Mycorrhizal association (Fungus).
- July-October is the most appropriate time for propagation.
- January end to the onset of monsoon is a critical period for irrigation of litchi.
- India ranks second in litchi production in the world.
- Wet spring, dry summer and light winter are desirable conditions for fruiting in litchi.
- Productivity in India – 7.51 MT/ha
- 75 per cent of the total area under litchi cultivation in the countryis in Bihar.
- 'Swarna Roopa' is an improved litchi cultivar selected in Bihar. Swaran Roopa is grown commercially in Chota Nagpur area. It is an early non-cracking seedless variety.
- Litchi mite is a serious pest in Kangra valley of H.P. while fruit cracking is the most serious physiological disorder of litchi.

  'Dehradun' cultivar is more prone to cracking of litchi than Calcuttia

Varieties

- Early Seedless – (Early benana) is a table & processing purpose variety.
- Calcuttia (Kolkatta) female flower (48.7%) fruit set 23.2%
- China, Elachi – Table purpose
- Shahi – suitable for canning/table purpose variety.
- Dehradun female flower (19.8%) fruit set (8.0%)

Other varieties

- Saharanpur, Kasha, Maclean, Purbi, Bombai, Desi, Glabi, Late seedless, Lath large red, Brewster, Maclean.

Composition

- Moisture – 77-83%
- Sugar – 6.74-18.86% in Indian
- Sugar – 12-15% in Florida
- Sugar – 11.8-20.6% in Hawaii
- Protein – 1.1%
- Acidity – 0.20 – 0.64%
- Vitamin B_2 – 122.5mg / 100g
- Calcium – 0.21%
- Phosphorus – 0.30%
- Juice – 60%
- Rag – 8%
- Seed – 19%
- Skin – 13%
- *Litchi philippinensis* is used as a rootstock for litchi.
- Dehradun, Calcuttia, Seedless Late, Rose Scented, Saharanpur and Muzaffarpur are commercial cultivars of Litchi.
- The cultivar 'Dehradun' is more susceptible to fruit cracking than cultivar 'Calcuttia'.
- Swarna Roopa is an improved variety of litchi evolves through selection.
- Inflorescence of litchi is a compound petalless raceme emerging from the terminal and axillary buds.

- Pollination of litchi is mediated through insects (honey bees and flies).
- Roots of litchi show association with mycorrhizal fungi in acidic soils. This association was reported by Coquille.
- Litchi flowers are staminate, hermaphrodite and pseudohermaphrodite (functionally male).
- The sequential opening of the flowers in litchi is male (Phase-I) followed by – Hermaphrodite (Phase-II) pseudohermaphrodites (Phase-III).
- Seedlessness in litchi is due to simulative parthenocarpy.
- Fruit cracking in litchi can be reduced by the spray of Boron @0.4%.
- Red rot of litchi is a disorder which can be controlled by a spray of lime sulphur.
- Litchi makes an excellent canned fruit.
- It was introduced to Myanmar and India towards the end of the 17th century and West Indies in the 18th century.
- Major Litchi growing countries – China, India, Taiwan, Vietnam, Thailand etc.
- Litchi is known by different names such as:
    - Thailand – Linchi, Australia – Lychee
    - Phillippines – Letsias, Malaysia – Laici
    - Indonesia – Litsi, South China – Laici
- The pH range of the soil for litchi cultivation should normally be 5.5-6.5.
- Litchi usuallyprefers low elevations but can be grown up to an altitude of 800 meters.
- Malic acid is the predominant nonvolatile acid is litchi.
- In China, about 20% of the total production is earned.
- Dried litchi, commonly known as litchi, is very popular in China.
- Litchi has originated in the region between Latin 23° and 27°N in southern china / northern china.
- Litchi was introduced in Myanmar and India at the end of the 17th century.
- Litchi is known as Linchi - Thailand Lychee – Australia
    - Liaïci – South China/North China
- Bengal is the second most important litchi growing region after Tai So in southern Queensland.
- Swarna Roopa is resistant to fruit cracking.

- pH range – 5.5-6.5
- The inflorescence is terminal of the branches while they are produced or hypoxanthine.
- Sabour Madhu (H-105) and Sabour Priya (H-73) are hybrids, resulting from (Purbi × Bedana)
- Chicken tongues – Litchi fruits that have aborted seeds are termed "Chicken tongues".
- Cracking early cultivars are more susceptible than late cultivars.
- Flowering starts from the last week of January or the first week of February and each year ripen in May-June.
- In South India, flowering starts in December and fruits ripen from April to May.

The ideal climate for Litchi

- Flushing – 28-30°C, high relative humidity, heavy rainfall.
- Dormancy – Min. 15°C, 50mm rainfall per month.
- Flowering – 16-20°C light rainfall.
- Fruit set – 18-24°C, moderate relative humidity.
- Harvest – 24-28°C even rainfall, High sunlight and high R.H.
- Pollination is optimum between 19-22°C.
- High temperatures (above 38°C) and R.H. below 60% results in severe skin cracking of fruit.
- Litchi seeds can be kept for 4 weeks in the fruit harvest.
- Layering – upright branches 1-1.5cm stem diameter and 30-60 cm long, done in June when temperatures are hot with high humidity and daily rain.
- Mulching – Mulching the tree basins at IIHR Bangalore resulted in better fruit size and simultaneous decline in fruit cracking.

Floral biology

- Panicles are 5-30 cm long.
- The duration of flowering (anthesis to pollination) usually ranges from 26-35days.
- Three types of flowers are commonly found in an inflorescence: male, hermaphrodite and pseudo-hermaphrodite (functional male).
- The flowers are self-sterile, requiring insects for pollination.

- Most fruit abscises in the first 2-4 weeks.
- Maximum fruit drop takes place during the first fortnight after fruit set and at harvest, the fruit retention varies between 3 to 39.6%.

Fruit drop

- Treatment for the same is done using growth regulators like NAA at 30-30ppm, GA_3 at 20-50ppm, 2,4-D at 10-20ppm, sprayed on panicles before the flowers open.
- The fruit maturity is optimal at 55 days after fruit set at the time of TSS-18.1°Brix, Acid-0.24%, S.G-1.00.
- The panicles should be cut from the tree with secateurs with as little as possible. Experiments have shown that removal of foliage reduces next season's flowering.
- Yield – Bears about 80 to 150kg fruits/year.
- Fruit for export is cooled immediately after harvesting to 0-2°C usually through cold water immersion or exposure to cooled air.
- Sulphur treatment is given before long-distance transport to increase the shelf life.
- Anthocyanin pigments-
    - Cyanidin – 3- glucoside
    - Pelargonidin –3- glucoside
    - Pelargonidin – 3 glucoside
- Storage
    - 20-25°C – 7 days
    - 4°C – 35 days
    - 1.7°C – 21-35 days
- Maturity time
    - I - Early - June
    - II - Middle - June
    - III - Early – July
- Dried litchi is known as 'litchi nut' and is popular in China.
- Leaves are used for making poultices and the seed as an anodyne for the skin.
- Fruit growth of litchi was described for the first time by Khan 1929.

Disease pest

- Litchi mite – (*Aceria litchi*) treatment involving application of sulphur fungicides to control this mite (karathane fungicide) is effective

Disorder

- Little leaf + leaf bronzing is caused due to Zn deficiency
- Fruit cracking is triggered by a combination of excessive water and high temperatures.

18

Plum

Botanical Name	-	*Musa paradisica* (Monocotyledoneus)
Botanical name	-	*Prunus domestica, Prunus salicina*
Chromosome No.	-	16, 32
Country of Origin	-	China
Fruit type	-	Drupe
Family	-	Rosaceous
Tree fruits	-	Deciduous
Bearing habit	-	Axillary bearing – Old season growth
Inflorescence	-	Cymose – Fascicle
Dichogamy	-	Protogyny
Pollination	-	Cross-pollination
Pollinates	-	Entomophilous
Fruit buds	-	Simple buds
Growth pattern	-	Double sigmoid growth
Respiratory behaviour	-	Climacteric
Rate of ethylene	-	High – 10-100 C2H4/kg/hr.
Storage life	-	4-8 weeks
Edible portion	-	Mesocarp and epicarp
Climatic	-	Temperate
Acid tolerance	-	Highly tolerant
Harvesting period	-	May-July
Area (India)	-	23.00 Thousand Ha
Production (India)	-	89.00 Thousand MT
Productivity (India)	-	3.86 MT/ha
Top producer in India	-	Uttarakhand> Himachal Pradesh> Jammu & Kashmir > Punjab> Nagaland

Top Producing in World - China> European Union> Romania> USA>Serbia

Storage - 0°C temp.

Propagation - Tongue grafting / T-budding

Spacing - 7.0 m × 7.0 m or 6.0 m × 6.0 m

Famous plum breeders - Burbank, California USA

- Fast track plum breeding started with an aim to reduce the juvenile period in temperate fruits.
- Transgenic early continuous flowering (ECF) gene induces early flowering.

Common Name	Origin
European - *P. domestica*	Europe (16)
Damson - *P. insititia*	Western Asia
Cherry or Myrobalan - *P. cerasifera*	Western Asia and Central.
Japanese - *P. salicina*	China – (48)
American - *P. americana*	North America
Simon - *P. simonii*	North America

Chilling requirement

- Japanese Plum - 700-1000 hrs - below – 7°C
- European plum - 1000-1200 hrs - 7°C
- Clonal rootstock for raising plum plants: Myrobalan B.
- Dwarfing rootstock – *P. subcordata*
- Planting time – Dec-Jan
- Training System – Open centre system – oldest
- Japanese plums are mostly adaptable to open centre system.
- Fruit thinning agents – DNOC, Ethephon, 3-CPA
- Japanese plum needs thinning upto 25 - 40%.
- European Plum: President, Victoria, Starking Delicious, Greengage.
- Japanese plum – Beauty, Santa Rosa, Mariposa. Kelsey, Frontier, Elephant Heart, Satsuma
- Subtropical – Sultlej Purple, Kala Amritsari, Titron.
- Self-fruitful – Beauty, Mariposa
- Self-unfruitful – Kelsey, Eldorado, Wickson Laredo and Formosa

- Popular – Frontiers, Santa Rosa
- Self-fruitful – 30% of flowers set fruits
- Self-unfruitful – 1.5% flowers set fruits
- Japanese plum varieties are mostly self-unfruitful and require pollination.
- Interspecific hybrids
  - Plumcot – Plum × Apricot
  - Plouts – Plum × Apricot × Plum
  - Apriums – Apricot × Plum × Apricot
  - Santa Rosa is a complex hybrid containing a mixture of *P. salicina* × *P. simonii* × *P. americana*
- Fruits TSS at maturity – 12.5% ^{O}Brix
- Plum pox virus (PPV) is a major problem that leads to heavy bearing.
- Sunburn is a major problem in tropical cultivars –.
- Breeding work started in the subtropical region in Saharanpur in 1957
- Deficiency of B in plum results in Misshapen fruits.
- Seedling rootstock – Zardalu (Wild apricot)
- In the market – Titron – May second week

  Jamuri – July thread week
- Excellent source of Vitamins A, calcium, magnesium, iron, potassium.
- Honey bees act as major pollinators
- Early varieties: Methley. Kelsey, Santra Rosa, Beauty, Settles, Cloth of Gold, Early Subza
- Mid – Satsuma, Elephant Heart, Frontier Victoria, Burbank
- Late – Mariposa, Red Ace, Late Yellow Grand Duke, Silver Wilkson
- Low hills – Alucha Purple, Titron, Satluj Purple, Alubakhara
- Alu Bokhara: Large fruit having yellow skin and red colour pulp which is juicy and sweet.
- It ranks next to peaches in terms of economic importance amongst the stone fruits.
- Prunes are a kind of plums with high sugar content above 18% which can be dried with a pit.
- In India, plum was introduced in Himachal Pradesh by Alexander Coutts in 1870

- Some low chilling varieties of plum were also introduced at PAU.
- Three flowers are produced in single bud on one year old shoot or the spur.
- Plum requires 90-110 cm well-distributed rainfall throughout the year.
- Soil pH – 5.5-6.5.

1. Prunus domestica (European plum)

It is a hybrid of diploid Myrobalan plum (*P. cerasifera* and tetraploid block thorn (*P. spinosa*).

The hexaploidy fruits are larger than Japanese plum.

The fruit is oval or round having both yellow and green ground colour and also both red and blue skin colour.

European plum is classified into three main groups:

a) Prunus: Fruit is oval with bulging ventral side and compressed bilaterally. It is blue or purple in colour, high in sugar content which makes it suitable for drying without removal of the pit.
   - All prunes are plums but all plums are not prunes.
   - Varieties are Italian Prunes, Giant Prune, President.

b) Reine Daude and Greengage plum: This is hybrid of *P. domestica* with *P. insitita*. The fruit is greenish yellow and round in shape having yellow skin and flesh. Important varieties are olden drop, green gage, Golden Transparent.

c) Lombard plum: The colour of the fruit is purplish-red. Lombard and Victoria are popular varieties.

2. Prunus salicina (Japanese plum)

- Originated in China but introduced in Japan from where it spread around the world. The plants are vigorous, productive, precocious and resistant to disease than the European plum. The fruits are large and heart-shaped with a pronounced apex. A few cultivars are oblate or round.
- *Prunus insititia*: This is a small-fruited European plum, hexaploid and grows wild in Europe and Western Asia. Plums of the species are known as Damson and Mirabelles. Fruits are small and purple (Damson) yellow (Mirabelles). Plant are small and compact and form excellent hedge rows.
- European plum cultivars: California Blue, Washington, French Prunes, Early Italian, Stanley, Grand Duke, Victoria, Damson.
- Japanese plum cultivars: Beauty, Methley, Santa Rosa, Kelsey, Mariposa, • Satsuma, Burbank Red Beaut, Frontier.

- American Cultivars: Destoto, Hauekeye Glyant, Weaver, Terry, Wayland, Golden Beauty.
- Clonal rootstock for plum
  - Myrobalan – B Mariana GF 8/1
  - Myrobalan – 29C Mariana 2624
  - St. Julian A Mariana 8-6
  - Pixy Mariana 9-25
  - GF-667
- Cuttings are taken from hardwood cutting and semi-hardwood is treated with IBA (2000-5000 ppm) for better rooting.
- Wild apricot is stratified under alternate layers of moist sand for 45-50 days at a temperature range of 3-5°C to break the rest.
- In Punjab, self-rooted plants of Kala Amritsari are generally used for planting.
- Seedling as well as clonal rootstock which is grafted in February buds in June-July.
- Most of the native American plums are self unfruitful, hence, need pollinizer.
- In case the cultivars produce little or no pollen, a pollinizer branch should be grafted on every tree.
- In plums, thinning and lea-ding back of shoots are two basic components of pruning.
- Most of the plum varieties bear on spurs on two years old wood.
- In bearing plum trees, 25-30% thinning per shoots and 50-75% heading back of shoots is suggested for proper fruiting.
- In rainy season, the weeds in the plum orchard are controlled with the post-emergence sprays of glyphosate at the rate of 800ml/ha or using mulching materials.
- Harvesting is carried out in two or three pickings.

Maturity standards of commercial varieties of plum.

Variety	Days from full bloom	Firmness	TSS (%)
Red beauty	88±3	5.60±0.5	15-16
Santo Rosa	94±3	5.90±0.5	16-17
Frontier	105±3	6.30±0.5	16-18
Kala Amritsari	85±3	5.40±0.5	13-14

Grading and packing of plum fruits

Grade	Fruit size Diameter (mm)	Box size	No of layers	No of fruit /Layer
Special	42 & above	14.5 ×6.5 × 6.5	3	28-32
Grade I	36-42	-do-	4	38-42
Grade II	below 36	-do-	4	50-56

- Plums being perishable have a very short shelf life.
- After evaluation, only Japanese plum has been recommended for commercial cultivation in the temperate region of north western Himalayas.
- Santa Rosa: Tree upright, vigorous and very productive. Fruit large, heart shaped, purplish crimson, Flesh amber in colour with reddishness near the skin. It is a soft fruitful variety with juicy and flavoured fruits.
- Frontier: Tree semi-vigorous, upright in growth and productive. Fruit large, skin purplish red, fruit heart shaped, flesh dead red, very sweet, juicy, firm and stone free. Fruit matures 10-15 days after that of Santa Rosa.
- Mariposa: An upright growing tree, fruit heart shape with greenish-yellow skin mottled with red. Flesh red in colour, juicy and firm. Late maturing variety.
- Kala Amritsari: This is a low chilling variety widely grown in plains. Self-fruitful. Fruit medium in size, round, oblate in shape, brownish in colour and turns dark drown at maturity, flesh yellowish, juicy with sub-acidic tastes.
- Chip budding during mid-February with smooth stock scion union is successful.
- T-budding during July-August is also recommended but the plant growth is poor.
- The best times of grafting is February for lower elevations and March at higher elevations.

FYM	- 60 kg/tree	year after 7 years
N	- 500g/tree	
P	- 250g/tree	
K	- 600g/tree	

- Fruit thinning chemicals for plum:

NAA	- 10-500 ppm
3-CPA	- 100-300 ppm
2AS-T	- 2-25 ppm
Sevin	- 1000-2500 ppm
DNOC	- 100-500 ppm
Paclo (PP333)	- 1000-2500 ppm

- The CA storage has been practiced overseas by maintaining 2-3% oxygen and 2-8% CO_2 and the fruits can be retained for a duration of 2-3 months.

19

Guava

Common Name	-	Apple of tropics
Botanical Name	-	*Psidium guajava* L.
Family	-	Myrtaceae
Chromosome No.	-	22, 33
Climatic adaptability	-	Tropical
Fruit morphology	-	Simple fruit (Berry)
Rate of Respiration	-	Climacteric
Photoperiodic responses	-	Day-neutral plant
Relative salt tolerance	-	Highly tolerant
Pollination	-	Cross pollination Insects (entomophilous)
Center of origin	-	Peru (South America)
Acid present	-	Ascorbic acid
Growth Curve	-	Double sigmoid curve
Area (India)	-	264.85 Thousand Ha
Production (India)	-	4053.51 Thousand MT
Productivity (India)	-	15.30 MT/ha
Top producer in India	-	Uttar Pradesh> Madhya Pradesh> Bihar> Andhra Pradesh> West Bengal
Top Productivity in India	-	Andhra Pradesh (24.12 MT/ha)> Punjab (22.50 MT/ha)> Assam (21.84 MT/ha)> Karnataka (19.58 MT/ha)
Top Producing in World	-	India> China> Thailand> Mexico>Indonesia
Top Productivity in World	-	Indonesia (17.95 MT/ha)
India Share and Rank	-	40.4% and first
Top Exporter (Dried/Fresh)	-	USA>UAE>Qatar>Nepal>Oman

Introduced in India by	-	Portuguese in the 17th century
Post-Harvest Losses	-	15.88%
Storage temperature	-	8-10°C
Propagation method	-	Stooling
Spacing (m^2)	-	7.0 m × 7.0 m (204 plant/ha)
Inflorescence	-	Solitary
Edible part	-	Thalamus & Pericarp
Ethylene production	-	Guava, Medium – (1-10) μlC_2H_4/kg/hr.
Bearing Habit	-	Axillary (Current season)
Type of Fruit bud	-	Mixed bud

- L-42 is the most suitable cultivar for canning.
- A promising genotype in Allahabad is locally known as "Lal Sebia".
- Strawberry Guava – *P. cattlcianun*, *P. nolle* is a shrub or small tree.
- Triploidy is the cause of seedlessness in guava.
- Red pulp colour is dominant to white and that this character is governed monogenic gene.
- Guava improvement work for the first time was initiated in 1907 at Ganeshkhind Experiment Station, Pune, Maharashtra.
- The fact that Guava wilt is soil-borne was first reported in 1935 from Babakkarpar Allahabad.
- Usually, fifteen days period is required for the complete wilting but some trees take even up to one year.
- In general, the more than 10-year-old plants are more prone to wilt incidence.
- Guava wilt, effects more severely in alkaline soils. Wilting is observed during the rainy season. It starts in August with the largest number of the plant dying in September and October.
- Guava improvement work was started at Pune in 1907.

Bahar	Flowering	Fruiting	Quality
Ambe	Feb-Mar	July-Sept	Watery poor
Mrig	June-July	Nov-Jan	Excellent
Hast	October	Feb-April	Good, yield low

- Fruit-bearing takes place 3 times a year in South India.
- Two types of fruits i.e., completely seedless and partly seeded are borne on the plant of the seedless variety.

- Guava is harvested throughout the year except May and June.
- HDP reduces TSS, sugars and ascorbic acid but increases titratable acidity
- Fruit quality of winter crop is best as it escapes the attack of fruit flies.
- The practice of taking winter crop instead of rainy season crop is known as crop regulation.
- Rainy season crop can be removed by the spraying of urea (10%) on Allahabad Sefeda and (20%) on L-49 at the time of peak flowering time, NAA – 400-800ppm, Ethiphon @1500-1500ppm in April-May.
- The state of U.P. produces the best quality guava.
- Vit. C content highest in fruit peel at a mature stage.
- Guava wilt is most commonly prevalent in alkali soil.
- Guava is a rich source of pectin.
- Guava is extremely beneficial for prevention of 'scurry disease'.
- Banding in Guava is practiced in Maharashtra.
- Dwarfing rootstock (Anueploid-82) Pusa Srijan
- Chinese Guava – *P. friedrichsthalianum* dwarfing rootstock and resistant to guava wilt and nematodes.
- Lucknow-49 (Sardar Guava)-A chance seedling selection from Allahabad Safeda developed in Pune in 1927 by Dr G.S. Cheema and Deshmukh (GFES). L-49 is more susceptible to bronzing than Allahabad Safeda, open-pollinated and tolerant to guava wilt.
- Chittidar – Fruits are characterized by numerous red dots on the skin.
- Harijha – Most popular variety in Bihar
- Hajsi – Red fleshed guava
- Apple colour – Tolerant of anthracnose
- Allahabad safeda – Parthenocarpic variety
- Behat coconut – Seedless guava
- Arka mridula – seedling selection from Allahabad Safeda, soft seeded variety
- Saharanpur Seedless and Nagpur Seedless
- Arka Amulya – Allahabad Safeda × Seedless
- Lalit – 24% higher yield than Allahabad Safeda; suitable for Jelly making.
- Shweta high TSS (24° Brix)
- Kohir Safed – Kohir × Allahabad Safeda

- Safed Jam – Allahabad Safeda × Kohir
- Meadow orcharding technique in guava was developed for horizontal utilization of space.
- Fruits of specific gravity between 1.00 to 1.02 SG and the best quality in the world produced in Allahabad region of India.
- Highly shade-tolerant crop
- *P. pumilum* – Most dwarfing
- Cross incompatibility between several cultivars limits the combining of superior traits in a single genotype (e.g., Lucknow-49 Apple colour and Behat coconut are not inter crossable).
- Seedlessness due to triploid genetic constitution
- Ganesh Khind fruit experiment station is located at Pune.
- H-16-1: Apple Colour × Allahabad Safeda
- Strawberry guava/Cattleya guava (*P. cattleianum*) is considered as an important species for use as rootstock.
- Brazilian guava /Quinea (*R. quineense*) has small fruit of poor quality.
- Mountain guava (*P. montanum*) is a shrub, about 1.5m high with flat branches.
- Fruit of *P. pomiferum* is apple shaped and that of *P. pyriferum* is pear-shaped.
- Guava is generally trained to modified leader system.
- Guava fruits can be stored for 4 weeks at 8-10°C and 80-90% humidity.
- Guava wilt is most common in alkali soils (pH 7.5-9.0).
- Since the fungus cannot grow at pH below 8.6, management of this disease involves keeping soil pH in this range.
- The Vitamin C content of fresh ripe fruit varies from 150 to 350 mg / 100 gm of pulp.
- The feijoa or pineapple guava (*Feijoa sellowina*) is grown on the hills of South India.
- The annual rainfall requirement is more than 250 cm (1,016 mm).
- It can grow in soils with pH ranging from 4.5 to 7.5.
- Some types are pear-shaped and some even look like a bitter gourd or Karela.
- Should be fertilized with 300-400g N, 250g P and 350g K along with 30-40kg FYM each year.

- Cracking of fruits occurs in extreme cases; spray of boric acid (0.3 to 0.4%) or 0.5%. Borax before flowering (July-August) is beneficial.
- Guava fruits ripen five months after flowering.
- 'Frank Malherbe' South Africa variety has been found to have 1,114 mg Vitamin C/100g of pulp.
- Guava plants are sensitive to waterlogging; hence, soil should be well-drained.
- The young plants are susceptible to drought and cold conditions. High temperatures at the time of fruit development causes fruit drop.
- Stooling – In this method, 3–5 year plants are cut back and allowed to shoot. Indole butyric acid (5,000ppm) in lanolin paste is applied in the previously made rings of shoots during July and soil is earthed up.
- It is better to restrict the tree canopy to a limited height in a rectangular shape, allowing more spread in "East-West" direction.
- In the early stages, plants require 8-10 irrigations a year.
- Crops like peas, cowpea, beans, gram can be grown as intercrops during initial stages of the orchard.
- About 80-86 per cent of flowers set fruits. But due to severe fruit drop, only 34 to 56 per cent of fruits attain maturity.
- Maximum flowers are dropped within 5-12 day of anthesis.
- Spraying of GA_3 at 15 to 30ppm in January is effective for increasing fruit retention.
- Grafted plants of guava start bearing fruits from the third year after planting. The fruit turns greenish-yellow with the advancement of maturity.
- Economic yield is obtained 7th- 8th year onwards.
- The average yield per tree is estimated at 60 to 90kg from the seedling trees
- 1000 fruits per year per tree can be harvested from a 10-year-old plant
- 500-600 fruit, per plant from the 7-year-old plant.
- Fruit fly (*Baclrocera dorsalis*) damages the fruits in the rainy season. The insects lay their eggs under the skin of the fruit.

  Control: Containing 100ml salutation of 0.1% Methyleuzenol and 0.1% Malathion.
- Scale insect is a major pest of guava in South India, MH, Punjab and UP.
- Bark eating caterpillar (*Indarbela* spp.) can be controlled by plugging the tunnels with 0.05% monocrotophos or 0.5%.

- Mealybug – (*Drosicha mangiferae*) has also been observed in winter season crop. The nymphs suck sap from leaves and young shoots. Chlorpyriphos dust (1.5%) @ 250g/tree may be applied by raking.
- Guava canker (*Diplodia netalensis*) is the second important fungal disease. Dithane Z-78 or Difoltan 0.2%.
- Wilt (*Fusarium oxysporium*) is sometimes encountered, especially in alkaline soils. It was first reported in 1935 from Babakkarpur (Allahabad) Bavistion (0.1%)

  Wilt in U.P – *F. oseysporum*

  Wilt in Bengal – *T. solani*
- Control of Anthracnose (*Glocosporium pridii*): Copper oxeychloride (0.3%) Difolatan (0.3%) at weekly intervals.
- Allahabad Safeda is more susceptible to this disease than Lucknow – 49 (L-49).

20

Grapes

Botanical Name	-	*Vitis vinifera*
Family	-	Vitaceae
Chromosome number	-	(2n) 38
Climatic adaptability	-	Sub-Tropical
Fruit morphology	-	Berry
Rate of respiration	-	Non-Climacteric
Pollination	-	Cross
Centre of origin	-	Egypt
Acid	-	Tartaric acid
Introduction Year	-	1300AD
Edible Part	-	Pericarp and Placentae
Storage tem.	-	16.0 °C
Storage tem.	-	0-1.0 3°C
Propagation method	-	Hardwood Cutting
Planting spacing	-	1.80 m × 2.40 m
Planting time	-	October
Optimum temp for growth	-	25-32°C
Area	-	139.00 Thousand Ha.
Production	-	2920 Thousand MT.
Productivity	-	21.02 MT/ha
Top Producicong States in India	-	Maharashta> Karnatka> Tamil Nadu> Mizoram> Andhra Pradesh
Top Productivity States	-	Punjab (28.67 MT/ha)> Tamil Nadu (27.27 MT/ha)>

in India Maharashtra (21.67 MT/ha)> Addhra Pradesh (20.00 MT/ha)> Karnatka (19.70 MT/ha)

Top Producting County In the World - China> Italy> USA> France> Spain

Top Productivity County - India (21.23 MT/ha)> China (17.60 MT/ha)> USA(17.31 MT/In the World ha)> Italy (12.28 MT/ha)> Chile (12.18 MT/ha)

Export of Grapes (Fresh) - Netherland> Russia> United Kingdom> Germany> United from India Arab Emirates

Post-harvest Losses - 8.63%

Parthenocarpy - Stermospermocarpy

Inflorescence - Panicale

Pollination - Wind and insects

Ethylene production - Very Low (<0.011-10) µl C2H4/kg/hr

Respiration rate - Low (5-10 mg) of CO_2 Release

Bearing habit - Current Season Grpwth

Flower bud differentiation - March-April (After 45 days of sprouting of buds)

- The skin of grape berry is covered with a wax-like layer which is called cutin
- Fe deficiency is very common in black soil
- Thompsom seedless variety with its done occupies 55 % area under grape cultivation
- Rains during ripening cause berry cracking and rotting
- Pruning
  - North India- December-January
  - South India- April and October
    - April- Back of foundation pruning
    - October- Fruit or forwarded pruning
- Mg deficiency is universal in grape cultivation
- CCC- for suppressing vigour of wine and increasing fruit fullness of bud
- GA3- For increasing berry size
- HCN- to hasten bud brack at winter pruning

- NAA- 50 ppm to reduce post-harvest drop
- MH- for induction of male sterility
- Bower system of training is mostly adopted in India (High economic ratio) cost-benefit ratio (1:2:9:)
- The average productivity of grapes in India (21.33 MT/ha) highest in the World
- Pruning intensity lowest (3-4 buds) in Bangalore Blue and Bophri
- Pruning intensity lowest (10-14 buds) in Thompson seedless (TS)
- The calyptra is a cap-like structure of grape formed as a result of a union of sepals and petals
- In saline soil dogridge rootstock is better than other rootstocks
- The highest mean bunch weight + berry weight is obtained under the bower system of training
- Berry drop in grape due to defective and improper pollination and fertilization
- Pink pigmentation develops in green grapes diurnal differences are more than 20 °C during ripening
- In nematodes-prone soil for 1612 rootstock can be used for Anab-a-shahi and TS
- For TS the spacing of 1.8 m × 3.0 m is ideal for the bower system
- Dipping berries in soda oil containing ethyl oleate + K_2Co_2 and shade during is most common method of preparing to raise in India
- Raising is only processed product in India
- Dogridge and 110 rootstocks are tolerant to drought and salinity
- Tartaric acid is commercially extracted from grapes
- Sub stance responsible for aroma in methyl anthranilate (muscat flavour)

  H.D. Olmo: Grape Breeder
- Queen of vineyard, Pusa Seedless are used as male parent to transmit seedless in common varieties fruit multi-seeded berry
- Maleic hydrazide is used for induction of male sterility
- Loose prelate from Prelate, Red Niagra from Niagra and Arabic Ordinal from Cardinal are grape mutants
- Bangalore Blu is a cross between *vinifera* and *labrusca*
- Delight and Perlatte ate sister seedlings of a cross between Queen of Vineyard and Sultanina Marble-26

- Salt creek belongs *V. champini* St. George belongs to *V. rucpestris*.
- Kali Shahebi is introduced in India under the name of Habsi from California
- Himrod grape is a cross between Ontario × Sultanina
- *Vitis parvifolia* is found in sub mountainous regions in India in functionally female due to reflected stamens
- The seeds constitute 0-10 percent of the weight of the berry and they are rich in tannin (5-8%) and cil (10-20%)
- The oil content in two of the Indian cultivars Anab-e- shahi and Kaliu Sahabi range from 10-13 %
- The grape berry generally contains four seeds
- The skin of the berries consists of a few layers of thick-walled cells called the epidermis. It accounts for 5-12 % of total berry weight
- The skin is covered with a thin layer of cutin and some times referred to as bloom.
- The cutin protects the berries from eater loss and attack of organisms and enhances the attractiveness of the fruits
- In the red grape, compound resveratrol has been found the check the spread of cancer
- The available information indicates a yield of 800-900 liters of juice per ten crushed berries

Vitis	
Equities Planch	Muscadinia Planch
Lenticels absent	Lenticels present
Diaphragm at nodes	Diaphragm absent
Long pyriform seeds	Simple tendrils, Ex With stout beak
V. vinifera	M(v) rotundifolia
V. labrusca	M(v)munsoniona
V. riparia	M(v)popenoei
V. rupestris	Note:- *Muscadine* is now
V. lincecumii and all other Vitis spp in North America	Considered separate genus characterized by mostly absence & presence of lenticels, simple tendrils and oblong seeds without a beak.

- *V. vinifera* CVs. – Anab-e-shahi, Bhokari, Cheema Sahebi, Gulbi, Kali Sahebi, Pandhari Sahebi, Phakadi, Sonaka, Tas, Thompson Seedless & Hobshi.

- *V. labrusca*: Bangalore blue, Bangalore Purple, Catauba, Concord, Isabella, Champion.
- *M. rotundifolia* CVs. – James
- *V. Chambini* CVs. – Dogridge, Chanpond Digraset.
- European grape (*Vitis vinifera*) origin – Caspier Black seas.
- *Euvitis* and *Muscadine* species originated in North America and were referred to as "Vineland".
- *Vitis vinifera* is, however, considered a hybrid between town American species *Vitis vulpina* L. and *V. labrusca* indicating that the Himalayan zone may be a seed centre of origin of *V. vinifera*.
- Anab-e-shahi is derived from the collections of Nawab Bagu Ali Khan.
- *Vitis vinifera*: (Wine grape) It produces the best quality fruits, the pulp is adhesive to the skin in American species. It is susceptible to most pests and diseases and most tolerant to salinity.
- *V. labrusca*: 2n-38. Black to amber in colour. The skin is thick and slips from the pulp at maturity is resistant to pests and disease.
- *V. aestivalis*: (2n=38) Pigeon grape, summer grape. Resistant to fungal disease, susceptible to phylloxera, adopted to the hot climate.
- *V. berlandieri* (winter grape) Spanish grape: Native to Mexican-38.
- *V. amurensis*: Resistant to cold and frost.
- *M. rotundifolia*: Resistant to phylloxera downy mildew. Native to the USA.
- India is considered a secondary place of origin of *V. vinifera* due to its resemblance with the characters of *V. paniuflar* and *V. lanata* found in the Himalayan region.
- Bokhari (syn Bhokri, Pocha Drakshi): It used to occupy almost 99% area in TN, MH and AP but it has now been replaced by TS and A-e-S.
- Rootstocks: Freedom, Harmony, Salt creek, Dogridge, St. Grarge etc.
- Dog ridge: Seedling selection from V. Champini suitable for light sandy soils and resistant to nematodes. The cuttings are difficult to root but budding and grafting are successful.
- St. George: (Rupestris St. George, Rupestris du lot). A seedling selection from *V. rupestris*, it is vigorous and resistant to phylloxera and drought and is easy to graft.
- Leleki 5A (*V. berlandieri* × *V. riparia*). It is highly resistant to phylloxera, tolerant to lime soils and moderately resistant to nematodes.

- 1613 (Couderc 1613, Soloais, Solanis x Othello). It shows moderate vigour to the session is resistant to nematodes and moderately resistant to phylloxera. It is easy to multiply by vegetative means.
- Selection 7 is a seedling selection from Pandhari Sahebi made by Dr Cheema now known as Cheema Sahehi.
- The north vines flower once during March-April, in Southern and Western region twice a year in June-July after April pruning and Nov-Dec after Oct punning.
- Fresh pollen grains are yellowish.
- The cultivar Bhakri has the largest pollen grains.
- A minimum of 6% fertility is sufficient for fruit set for successful hybridization programs.
- Nalwadi *et. al.* (1972) observed maximum pollen for the utility of 99% in Anab-e- Shahi and a minimum of 71.1% in Kandhari.
- Stratified at 4°C for 75 to 90 days before sowing to break the dormancy.
- Dormancy in grapes seeds is mainly attributed to the presence of ABA like growth inhibitors.

21

Peach

Common Name	-	Aadu
Botanical Name	-	*Prunus persica*
Family	-	Rosaceae
Chromosome No.	-	16
Origin	-	China
Fruit type	-	Drupe
Tree fruits	-	Deciduous
Axillary bearing habit:	-	Old season growth - one year Inflorescence - Racemose-solitary
Pollination	-	Self-pollination
Pollinator	-	Entomophilous
Growth pattern	-	Double sigmoid growth
Acid tolerant	-	Slightly tolerant
Respiratory	-	Climacteric
Rate of respiration	-	Medium – 10-20mg of Co2 kg/ha.
Storage life	-	4-8 weeks.
Edible part of fruit	-	Mesocarp
Major colour compounds	-	Orange-yellow
Pigments	-	Flavonoids
Genome size	-	220-230mbp (2010)
Climatic adaptability	-	Temperate
Harvesting period	-	May-July
Field	-	7-10MT/ha
Storage	-	0-0.3°C
Propagation	-	T-budding/Tongue/Cleft grafting.
Spacing	-	5 m × 5 m
Fruit bud	-	Simple bud

- Highly susceptible to waterlogging.
- Self-compatible in nature.
- Low chilling peach - <500 hrs chilling requirement.
- Red blush skins are caused due to anthocyanins.
- The yellow colour of peach is due to xanthophylls.
- Acid present - Malic acid
- Prunasin is the principal glycoside present in the pulp.
- Amygdalin is present in the seeds of peach.
- The oil content of peach seeds is 40-50%
- *Prunus behmi* is a natural hybrid of Almond and Peach.
- - Smooth skinned peaches (*Prunus persica* var. nucipersica) also called nectarine.
- Yellow fleshed peaches are ideal for table and canning purpose.
- white-fleshed peach is ideal for dehydration purpose.
- Florida sun, Sharbati, Saharanpur Prabhat are yellow fleshed peach.
- Nectarine is a fizzler (smooth-skinned) peach with strong flavours and aroma.
- Nectarine is a single dominant gene mutation in peach for fuzziness.
- Training system - Open centre system
- Peach requires regular and heavy pruning.
- Appropriate pruning time is from December to January
- HDP – 3 m × 3 m
- Tatra trellis system – 5 m × 1 m – 2000pl/ha
- Meadow system – 2 m × 1 m (5000pl/ha)
- Spraying of GA3 @ 200ppm or Ethephon is effective in countering the spring frost which leads to delaying of the bloom period -.
- Nectarine cultivars - Nectared, Sun Grand, Sunlight, Sun Red, Sunrise, Sun Ripe, Summer Queen, Red Gold, Silver King, May Fire, Independence
- Midseason variety - Early grande
- Early ripening - Florida Prince
- Low chilling varieties - Saharanpur Prabhat. Floundered, Sun Red, Sun Gold, Share-e-Punjab, Sharbati,
- Male sterile variety - J H. Hale
- Sharbati - Developed through clonal selection.

Low chilling Varieties: Shan-e-Punjab, Saharanpur Prabhat, Pratap, Florida Prince, Sharbati, Punjab Nectarine, Early Grande

Natural hybrid

GF-557, GF-677 (Peach × Almond) – Widely used rootstocks for peach and almond.

Interspecific hybrids

- *Prunus halsiana* × Yunman – Tolerant to drought and root-knot nematode, Armlaria, verticillium disease.
- Marianna/ Damas - *P. domestica* × *P. munsoniana* preferred rootstock for heavy soil.
- Nemaguard - *P. persica* × *P. davidiana* widely used as a root-knot nematode-resistant rootstock.
- Redhar - Halehaven × Kalhaven
- Prabhat - Sharbati × Florida Sun
- Regal - Resistant to cold and bacterial spot disease.
- Recent hybrids - Gala, Glory
- Nematodes resistant rootstock - Nemaguard, Nemared, Shaline and Yunnan.
- Nectarine cultivar is mostly preferred for table purpose.
- Peach leaf curl is caused by *Taphrina deformans* (fungus).
- Peach leaf cult aphid (*Brachycaudus helichrysi*) is the most serious pest of peach.
- Peach short life syndrome (PSLS) is caused by ring nematode.
- Physiological disorder - splitting and gumming is due to prolonged dry period and sudden rain.
- Splitting of fruits (at pit hardening stage) and gumming cause: unknown or undetermined.
- *Prunus behmi* is a natural hybrid between Almond × Peach.
- TSS about 8-13°Brix
- Swelling buds are injured at 6.5°C
- Peach is very susceptible to Fe deficiency.
- Peach requires high proportions of N and K.
- Application of Ethephon (300ppm) within 20 days after petal fall in Julie Alberta is recommended for optimum fruit thinning.
- Bears fruit after 2years.
- It has high nutritive value since it is rich in proteins, essentials amino acids minerals and vitamins.

- Peaches were introduced in the mid-hill regions of the state of Himachal Pradesh by Mr. Alexader Coutts in 1870 on the advice of an American horticulturist named Prof. R.W. Hodgson.
- Some low chilling Californian varieties were introduced at PAU during 1968.
- Peach is a temperate zone plant and its commercial production is confined to latitudes between 30 and 40°N and S, although it is now being grown almost all over the world.
- The leaf is simple, large, oblong-lanceolate, glabrous above and pubescent beneath.
- Vegetative and flower buds are borne in the axil of leaves.
- The flowers are numerous and sessile, having white or pink colouration and appear before leaves.
- Peaches require a humid climate with cold winters and dry summers.
- Peaches need about 500 to 800 chilling hours during winter months to break bud dormancy.
- Chilling requirement is easily met in areas located at altitudes between 1000m to 1600 m above mean sea level.
- Low chilling varieties require 250-300 chilling hours.
- pH – 5.8-6.8
- Peach seedling is generally used as a rootstock for plum. Apricot and almond seedlings can be also be used.
- In the plains, seeds of Sharbati are used as rootstock.
- GF677 is useful on alkaline soil due to resistance to chlorosis.
- Important clonal rootstock – Siberian – C (Cold hardy)
  St. Julien hybrid No. 1 and 2.
- Before sowing, seeds are first stratified at 4-5°C or below for 10-12 weeks in moist sand.
- Seed pre-sowing treatment – Thiourea-0.5% GA-200ppm, BA-100ppm.
- The clonal rootstock is multiplied through the mound and trench layering.
- In the hills, tongue grafting is done during February and T budding during May-June.
- In plains, grafting period is November to January.
- Budding happens during the months of April-June and in September.
- Manure and fertilizer – FYM-10, NPK-70:35:100g/year after 7 year & above – FYM-40 kg NPK-500:250:700g/tree.

- Weedicides like simazine and atrazine –
    - Atrazine - 2.0 kg/ha Pre-emergence
    - Terbacil - 0.8 kg/ha Pre-emergence
    - Paraquat - 4.0 litre/ha Post-emergence
    - Glyphosate - 4.32 kg/ha Post-emergence
- Application of Ethephon 300 ppm is recommended during petal fall in July
- Elberta is recommended for optimum fruit thinning.
- Ethephon (800 ppm) should be used for thinning 20-30 days after fruit set when the fruits are 20-25 mm in diameter.
- 78 to 127 days are required from flowering to maturity; the number varies across different cultivars
    - Flordasum - 81days
    - Alexander - 86 days
    - July Elberta - 101 days
    - Elberta - 127 days
- Peach leaf curl aphid (*Brachycaudus helichrysi*)
    - Major attack in - March
    - Control - 0.025% methyl demeton
      - 0.03% dimethoate in 200 litres of water - 7-10 days before flowering.
- San Jose Scale (*Quadraspidiotus perniciosus*) Current season growth.
    - Servo oil - 4 litres/200L water in Feb.
    - Metasystox - 200 mL-200L of water after harvesting of fruits.
- Peach fruit fly (*Dacus* spp.) - April and May 0.1% malathion and 1% sugar.
- Defoliating beetles - 1 kg carbaryl - 500 litres of water.

Disease

1. Peach leaf curl: (*Taphrina deformans*)
    - Copper oxychloride – 300g/100L of water)
2. Bacterial gummosis: (P*seudomonas syringae*)
    - Streptocycline – 10g/100L of water

    Mashobra paste copper oxychloride – 300g/100L water
3. Bacterial spot (*Xanthomonas pruni*)
    - Copper oxychloride – 300g/200L of water

22

Kiwifruit

Common Name	-	Chinese gooseberry, China's miracle fruit, Horticulture, the wonder of New Zealand / National symbol of New Zealand.
Botanical Name	-	*Actinidia deliciosa*
Family	-	Actinidiaceae
Chromosome No.	-	58
Origin	-	Southern China
Type of fruit	-	Berry
Climate	-	Temperate
Edible portion	-	Endocarp
Pollination	-	Cross-pollination
Pollination	-	Dioecious
Growth pattern	-	Triploid sigmoid growth
Acid tolerant	-	Slightly tolerant
Respiratory	-	Climacteric
Ploidy levels	-	Auto-hexaploid
Flowers are borne	-	Current season shoots
Leading producer of Kiwi	-	Italy
Flower colour	-	Creamy white to yellowish
Total phenotypic gender expression	-	6

- Gender change due to bud mutation in the mature male vine.
- *Actinidia arguata is acold tolerant species.*
- Commercial propagation - Stem cutting, Softwood cutting
- Flowering time is from the last week of April to 3rd week of May.
- Storage rot, a major disease during storage is caused by *Botrytis cinerea*

- Hayward is one of the a most popular cultivars in the world.
- It is a deciduous wine.
- It was planted for the first time in the Lal Bagh garden at Bangalore around 1960.
- T-bar or Pergola is adopted for training the vines.
- It is a rich source of Vitamin-C.
- The skin of the fruit is a good source of flavoured antioxidant.
- Kiwi fruit is also used for curing cancer.
- Crown gall is the major disease found in this fruit.
- The kiwifruit (*Actinidia deliciosa*) (Chev.) is a deciduous fruiting vine native to Yangtze River valley of South and Central China.
- After 1960, its cultivation gained momentum in other countries of the world, and now it is cultivated on a commercial scale in the USA, Italy, China, Japan, France, Germany and Australia.
- It was introduced in Shimla hills in 1963, where the plant came into bearing in 1969.
- It is rich in Vitamin-C, potassium, phosphorus and iron and is low in calories.
- Soil pH of– 5.5-6.5 is considered ideal for vine growth and fruit production.
- Can be grown successfully between 3000-5500 feet amsl which provides 600-800 chilling hours to break dormancy.
- Low temperature (-2.5°C) and frost during spring and autumn is very harmful for the fruit as it kills immature shoots and fruit buds.
- Summer high temperature >38°C accompanied by high insulation (100) and low humidity may cause scorching of leaves and sunburn of fruits which results in the death of the plants.
- Well distributed rainfall of about 120-150cm throughout the growing period is sufficient for proper growth and development of the fruit.
- Pistillate varieties -Hayward, Allison, Abbott, Monty, Bruno.
- Staminate varieties: Allison, Tomuri, Matua.
- Seedlings of some cultivated varieties like Bruno and Abbott are commonly used as a rootstock.
- Shoot is cut during January-February for making hardwood cutting.
- 4000-5000ppm IBA solution for 10-12 seconds.
- Before sowing, seeds are stratified in permanent layers of moist sand for 30- 35 days at 0-5°C to break dormancy.

- At the three-leaf stage, the seedling is transplanted in polybags and than in nursery beds during July-August.
- Tongue grafting and chip budding is done in last week of January to end of February that give 90-95% bud take success.
- Planting distance 6.0 m × 4.0 m planting- Allison, Abbott, Monty –trained on T-bar trellis system and 5.0 m × 6.0 m planting- Hayward
- Female adjacent to male is achieved by 1:8 or 1:9 male to female ratio
- In the T-bar trellis system, the pillars of iron and concrete are erected at about 1.8m height above the ground level.
- The fruit is developed on current season's growth arising from one year shoot.
- Only the basal buds of the modes 4-12 on current season growth are productive.
- Vines grow 2-3m long every year
- During summers, the pruning shoot is cut beyond 6-8 buds from the char fruit during June–July.
- About 8-9 colonies of honey bees are found per hectare.
- Besides, insect pollination and hand pollination are essential for getting fruits of better size and quality.
- For fully bearing vines (8years and above) 60-80kg FYM, 800g-N, 560g P, 1200g K vine is applied every year per vine.
- N fertilizer is applied in two equal dressing, half before bud-burst and the remaining full dose at the onset of monsoon *i.e.*, in July.
- Water requirement of kiwifruit is very high because of vigorous vegetative growth and large leaf surface area.
- Fully-grown vines require 80-100 L of water for total daily transpiration from 16-17m2 canopy area during summer.
- Drip irrigation at 100% ETC produces a higher yield of quality fruits.
- In hand thinning only lateral flowers or fruits are removed.

Varieties	DFFB to harvest (Lower elevation)	Harvesting Dates	DFFB to harvest (higher elevation)	Harvest Date
Allison	193±4	20-25 Oct	209±4	4-6 Nov
Abbott	190±4	20-25 Oct	211±4	8-12 Nov
Bruno	182±4	15-20 Oct	205±4	1-3 Nov
Monty	192±4	20-25 Oct	-	6-10 Nov
Hayward	201±4	1-5 Nov	223±4	18-20 Nov

- In India grades –
    - 'A' grade - >70g
    - 'B' grade - 50-70g
    - 'C' grade - <50g
- International grades –
    - 'A' grade - >100g
    - 'B' grade - 70-100g
    - 'C' grade - <70g.
- In India, Kiwifruit is packed in 1 Kg CFB packs or even polyethylene grooved trays.
- In a tray, 33 fruits of 100g/fruits weight are accommodated.
- In can be stored for 4-6 months in cold storage at 0°C and 90% RH.
- A large number of processed products such as jam, jelly, candy, squash and wine are prepared from Kiwifruit.

23

Apricot

Common Name	-	Coppery fruit
Botanical Name	-	*Prunus armeniaca*
Family	-	Rosaceae
Chromosome No.	-	16
Origin	-	North-Eastern China
Fruit type	-	Drupe
Tree fruits	-	Deciduous
Climate	-	Temperate
Inflorescence	-	Racemose – Solitary Self-pollination (Autogamous) - Homogamy
Self-incompatibility	-	Gametophytic
Fruit bud	-	Simple bud
Growth pattern	-	Double sigmoid growth
Respiratory	-	Climacteric
Storage life	-	0-4 weeks
Edible part	-	Mesocarp and Endocarp
Harvesting period	-	May-June
Yield	-	15-22 (MT/ha)
Storage	-	0°C
Propagation	-	Tongue grafting
Spacing (m)	-	6.0 m × 6.0 m

- Drought resistant, Salt tolerant and hardy temperate fruit crop
- Highly perishable fruit
- Vitamin A 2600IU
- Thiamine : 217mg/100g
- Wild apricot or Zardalu has originated from India.

- Popular training system – Open vase system
- Apricot bears on spurs and laterally on 1year-old shoots.
- Apricot spur life – 3-4 years
- Ideal thinning agent – NAA @25-50ppm, 20 days after fruit set
- Ideal stage for freezing, canning and drying fully ripe fruits
- Chaubattia Alankar – Kaishu × Charmagz – Low chilling and early ripening
- Chaubattia Madhu – Turkey × Charmagz – Early ripening
- Chaubattia Kesri – St. Amrbois × Charmagz – mid season variety
- Low chilling subtropical varieties – New Castle, Early Shipley, St. Ambroise, Kaisha.
- Important varieties – New Castle, Kaisha, Charmagz, Nugget, Moorpark.
- It is rich source of pectin and oil (40-45%).
- Moor Park– One of the best apricot varieties for outdoor cultivation in a small garden.
- Largest producer – Turkey
- Chilling requirement – 300-900 hrs below 7°C.
- Peak water use period – April end to Mid-June.
- Summer temperature – 16.6-32.2°C
- Used for treatment of tumors and ulcers.
- High contents of antioxidant (Carotenoids)
- Sweet kernels are used in confectionery while bitter kernels are used for oil extraction and raising rootstocks.
- The apricot fruit moved westward from Central Asia through Iron and Transcaucasus region and reached Italy during the first century, England in the 13th century and North America by 1720.
- Cultivation of Commercial cultivars of apricot in India was started recently by European settler and missionaries after 1870.
- Early varieties: EMA, New Castle, Early Shipley, Kaisa, Nugget, Royal St. Anbrosis, Moorpark.
- Late varieties: Shakarpara, Suffaida, Chamagz, Nari
- Dry Temperate: Charmagz, Suffaida, Shakarpara, Kaisha, Halman, Nari, Tokpop
- New varieties for mid-hills
    - Early maturing – Baiti, Beladi, early maturing apricot
    - Late – Farmingdale, Alfred

- White fleshed sweet kernelled apricots require cooler climate and are grown in the dry temperate region up to 3000m amsl.
- Spring frost cause extensive damage to the blossoms which are killed when the temperature falls below 4°C.
- Soil pH – 6.0-6.8
- Different clonal rootstock
  - Myrobalan 29C California Mariana 2642
  - GF31, GF8 – France
  - Mariana 7/7 – South Africa
- Seeds are stratified for 45-50 days at 4°C to break dormancy.
- Chip budding performed in September give successful results.
- A density of 7200 trees/ha has been reported in the cv. Canino
- Most of the commercial varieties of apricot are self-fruitful and set fruits without pollinizer.
- Charming and Perfection have been reported to be self-incompatible.
- There is 40-60% fruit set in the cultivars commercially grown in mid-hills, but fruit crop is to the extent of 79% in their cultivars, which occurs mostly in the second week after fruit set.
- A spray of 10ppm NAA at the beginning of pit hardening reduces the pre-harvest drop.

	FYM	CAN(g)	N(g)	SSP(g)	P(g)	MOP(g)	K2O(g)
One year	10	280	70	220	35	165	100
7 and above	40	2000	500	1560	250	1170	700

Name of Variety	DFFB to harvest	TSS° Brix
Shipley's Early	73±4 days	15-16
New Castle	84±4	12-15
Blenheim	85±4	15-16
Shakapara	96±4	14-17
Royal	100±4	12.5-15
Kaisha	92±4	12-15
Alfred	86±4	13-16
Baiti	82±4	11-13
Beladi	81±4	12-14
Farmingdale	84±4	12-15

- The apricot tree starts fruiting at the age of 5 years and gives economic yield up to 30-35 years.
- Apricot attains full bearing age at about 8-10 years and yields about 50-80 kg fruit/tree.

Grade	Fruit size	No of layers	No of fruit/layer	Size of box (cm)
Special	42 and above	3	28-32	37×16.5×16
Grade I	36-42	4	38-43	37×16.5×16
Grade II	L36	4	50-56	-do-

24

Cherry

Common Name	-	Sweet cherry
Botanical Name	-	*Prunus avium*
Family	-	Rosaceae
Chromosome No.	-	16
Climate	-	Temperate
Origin	-	Asia Minor
Tree fruits	-	Deciduous
Self-incompatibility	-	Gametophytic
Pollination	-	Cross-pollinations
Pollination	-	Entomophilous
Fruit bud	-	Simple bud
Growth pattern	-	Double sigmoid growth
Respiratory	-	Non-climactic
Storage life	-	0-4 weeks
Edible part	-	Mesocarp and epicarp
Barbados cherry	-	Vitamin C (1000-4000 mg/100g)

- Native to temperate regions of Northern Hemisphere.
- Chilling requirement – 2000-2700 hr, highest among temperate fruits.
- Heavy rainfall during flowering causes – Blossom wilt.
- Heavy rainfall during ripening causes fruit cracking.
- In Europe, a wine – 'Kirschwasser' is distilled from the pulp of cherries.
- Most of the commercial varieties of cherry are self-sterile.
- Seedling rootstock – Pazza, Mahaleb, Mazzard
- Clonal rootstock – Colt, Mazzard, F-12/1
- Cherry fruits are the first to appear in the market among temperate fruits.

- Cherry has more calorific value them apple.
- Donar vars: Stella, Vista, Vic, Seneca, Vega.
- Trained by a modified leader system.
- Cherry is grown under rainfed conditions in India.
- The red pigment in cherries is Anthocyanins (antioxidant)
- Cherry is sensitive to frost and water logging.
- The flavour of cherry is due to the presence of Methyl anthranilate and methyl salicylate.
- Colouring compound present in cherry – Keracyanin chloride.
- World-leading breeding centre of cherry: John Innes Horticultural Institution, Merton.
- Sweet Cherry – *P. avium* – Mostly used for table purpose.
- Sour Cherry – *P. cerasus* – Mostly used for canning or cooking.
- Duke Cherry – *P. gudowini* – Interspecific hybrid.
- Paja (*P. padus*) – Delayed incompatibility
- Wild bird cherry – (*P. padus*)
- Mahalab – (*P. mahaleb*)
- Mahleb – is also known as St. Lucie cherry.
- Clonal rootstocks – Colt and Mazzard F-12/1
- Cherry cordial – Monoschinos
- Abnormal reduction in fruit size – Zinc
- Interveinal chlorosis – Mg.
- Sweet cherry varieties – Early Richmond and Montmorency.
- Resistant to fruit cracking – Sam, Sue, Windsor, Victor.
- The cultivated cherries are divided into two main groups *i.e.* sweet cherries (*P. avium*) and sour cherries (*P. cerasus*).
- Sweet cherry is mainly used for table purpose and sour cherries for processing.
- Cherries are rich in protein, sugar, potassium, calcium, iron and zinc.
- The earliest records indicate that cherry was first domesticated in Greece around 300BC.
- Cherry spread to Italy from Greece where it was established as a fruit crop in 37 BC.

- Early settlers brought its seeds to North America and part of South America.
- In India, cherry was introduced by British settlers in Kashmir, Kullu and Shimla hills during the pre-independence era.
- Sweet cherry is a tall tree; branches are erect, leaves are large, thin, pubescent beneath and serrated.
- Only lateral, simple flower buds are borne on 2 years old shoot or at the base of 1-year-old shoot.
- Flowers are white and have 5 petals, numerous stamens, single style and an ovary with a single carpel containing two ovules.
- Sweet cherries have been divided into two pomological groups:

  a) Heart Group

  b) Bigaredee Group
- Heart Group: The heart cherry varieties have soft and tender flesh and heart-shaped fruits. The fruit colour varies from dark with reddish juice to light coloured with colourless juice.

  Varieties: Red Hearts & Black Hearts.
- Bigarreau Group: The Bigarreau Group of cherries is usually roundish. The colour of fruit and juice also varies from dark red to light red.

  Varieties: Sam, Summit, Sue, Sunburst, Lapins, Compact Stella and hybrid (13-17-40).
- Sweet cherry requires cool climate
- It is grown successfully at altitudes between 2000-2500 m amsl pH 6.5-7.0. Soil that can hold moisture during summer is most suitable.
- Cherry plant is very sensitive to water logging and therefore heavy soil should be avoided.

Varieties

Jammu and Kashmir: Black Heart, Early Purple Heart, Guigne Noir, Cross Lucent, Guigne Noir Hative, Guigne Pour Dua, Precece, Biganeau Napoleon Biganeau Noir Gross.

Himachal Pradesh: Black Tartarian, Bing, Napoleon, Sam Sue, Stella, Van, Lambert, Black Republican, Pink Early, Black Heart, Early River, Sunburst, Duro-Nuro II and Merchant.

Uttarakhand: Bedfored Prolific, Black Heart and Governor's Wood.

- The seedling rootstock of cherry is Paja (*P. paddum*), bird cherry (*P. paddum*), Mahatab and Mazzard.

- Clonal rootstock is colt, Gisela, Charger, SL 64 and Mazzard F12/1.
- Mazzard and F12/1 rootstocks are semi-vigorous and difficult to root.
- Colt is semi-dwarf variety *i.e.* compatible with almost all varieties of sweet cherry, has good anchorage, and is tolerant to gummosis, crown-rot, moderately resistant to stem pitting virus and bacterial canker but susceptible to Oak- root fungus.
- Seeds are soaked in 500ppm GA3 for about 24hrs. Then they are stratified by placing between the layers of sand in a cool place at 2-4°C for 80-120 days for mahaleb and 120-150 day for mazzard to break seed dormancy.
- It difficult to root mazzard F12/1 rootstock, IBA (7500ppm) is applied to the ringed portion of the shoots during summer.
- Colt rootstock is easy to root and can also be multiplied through cuttings.
- Cuttings are treated with IBA (2500ppm) for 10 seconds and planted in nursery buds for rooting.
- Tongue grafting during February-March is recommended for obtaining a bud take of more than 90%.
- For semi-dwarfing rootstock like colt, spacing can be planted 4.5m × 4.5m.
- Most of the cherry varieties are self-incompatible as well as cross incompatible.
- Weed central
- Diuron – 4kg/ha – pre-emergence
- Paraquat – 0.5% post-emergence.
- Fruit cracking is a serious problem in cherry which causes 50 to 80% losses.
- The spray of calcium chloride at 300g per 200 litre water or GA3 at 2000 ppm or NAA 10 ppm at 25-30 days before harvest checks fruit cracking.
- The fruits are packed in boxes lined with paper. Nowadays, 1kg cardboard boxes are used for packing.

25

Avocado

Common Name	-	Butter fruit/fruit of new world
Botanical Name	-	*Fersea americana*
Family	-	Lauraceous
Chromosome No.	-	24
Climatic adaptability	-	Subtropical
Fruit morphology	-	Berry
Relative acid tolerance	-	Medium tolerant
Type of dichogamy	-	PDSD
Type of pollination	-	Goss
Harvesting time	-	Aug-Sept
Yield (t/ha)	-	7.5 to 11.00
Propagation by	-	Air layering / T. budding
Ethylene production	-	High (10-100 µl C2H4/kg/hr)
Respiration rate	-	High (20-40 mg)
Origin	-	Mexico / Southern Mexico

- Avocado is a dense and evergreen tree can grow about 80 feet height
  - Single seeded berry - Fat – 24-26%
  - Recalibrate seeds - Sugar – 1% Rich in K, Fe, Vitamins – B
- Its energy value is twice as much as Banana fruit.
- Most leading avocado cultivar in the world – Fuertes.
- Avocado is classified into 3 distinct horticultural races.
  - i) Mexican – Coli fried Duke Pernod
  - ii) West Indian – Pollock, Simmonds, Black Prince Fuchsia, Peterson, Waldin.
  - iii) Guatemalan – Taylor, Linda, Queen, Benit
- Duke seedlings are resistant to root rot and cold hardiness.

- Important to salad fruit
- Pollock – can overcome the salinity problem.
- Girdling of alternate bearing varieties increases the yield.
- Indian varieties: -
  - i) Green (Guatemalan type) oval shape fruit.
  - ii) Purple (West Indian type) pear shape fruit.
- In India, it is grown in the slopes of Nilgiris coastal region of Karnataka, Kerala and Maharashtra.
- The fruit is harvested in Aug-Sept in South India.
- Commonly recommended spacing of avocado – 7.0 m × 7.0 m.
- Since the fruit contains not more than 1% sugar, Avocado is recommended as high energy food for diabetics.
- It takes of avocado cholesterol levels in the human body.
- Fuerte – Mexican × Guatemalan, alternate bearer, pear-shaped, weight – 400 g.
- Edible portion – Pericarp
- The pulp has a buttery texture, a rich nutty flavour and oil up to 26%.

Rootstocks

- West Indian avocado – Salt tolerant
- Mexican avocado – Tolerant to phytophthora root rot, Verticillium wilt, Dothiorella canker.
- Avocado can be grown at an altitude of 600-1500 m.
- High humidity during flowering and fruit set is necessary to secure a good crop.
- Harvesting index: oil content
- Mexican race and its hybrid are well adapted to the cool climates.
- West Indian types are best adapted to the law land tropical conditions and humidity.
- Guatemalan race is intermediate.
- Loamy or sandy routs of alluvial origin 5-7pt.
- Avocado is sensitive to waterlogged condition.
- West Indian race is tolerant to salinity.
- Avocado has more than 400 varieties and 3 distinct races.

- West Indian race matures in 60 months, has large seed & lose cavity
- Guatemalan fruit matures in 9 months, has smaller seed & tight cavity.
- West Indian cultivars are generally used as fruits.
- Guatemalan possesses coarsely granular skin.
- The most leading avocado cultivar in the world is Fuertes, (Mexican × Guatemalan) hybrid pear-shaped fruit, weight – 400g/fruit.
- The pulp has a buttery texture, a rich misty flavour and contains oil up to 26%.
- In parts of South India and Maharashtra, two varieties are cultivated namely Purple (West India Race) and Green (Guatemalan Race).
- Purple variety has pear-shaped fruits and weight of 450g/fruit.
- Single trees of avocado are not productive at a time for want of pollination, Hence, while raising a plantation in the new area, mixed planting of cultivar is desired instead of monoclonal stands.
- The avocado trees show water stress suddenly by shedding fruits and leaves or by wilting as they have a slow root system.
- Most resistant to cold temperature.
- The fruit is a rich source of oil ranges from 5-20%. Used for the cosmetics industry.
- The optimum temperature for flower induction: <25°C
- Suitable for the low-temperature condition: Mexican race
- Type of inflorescence compound panicle of the raceme
- Mode of pollination: Honey bees
- Viability of seeds: 2-3 weeks
- Frost resistant rootstock: Mexican types
- Protogynous diurnally synchronous dichogamy (PDSD) was 1st reported by Bergh (1969).

Particulars	**Mexican Races (*P. americana* var. drymifolia)**	**Guatemalan Races (*P. americana* var. guatemalensis)**	**West India Races (*P. americana* var. americana)**
Climate	Semi-tropical	Subtropical	Tropical
Cold Tolt.	High	Medium	Low
Salt Tolt.	Low	Medium	High
Oil	30%	8-15%	3-10%
Mature	6 months	>12 months	5 months
Varieties	Duke, Topa	hula, Hass, Green	Pollock, Purple

- The fruit can be kept well for 2 months at 0°C for one month at 4.5°C and for 15 days at room temperature.
- Fruit cracking in pomegranate of Boron defiant.

26

Cashew

Botanical Name	-	*Musa paradisica* (Monocotyledoneus)
Common Name	-	Dollar earning crop, Plough crop, Gold mine of the wasteland.
Botanical Name	-	*Anacardium occidentale.*
Family	-	Anacardiaceae
Chromosome No.	-	42, X = 21
Climatic	-	Tropical
Fruit morphology	-	Nut
Rate of respiration	-	Non-climacteric
Salt tolerance	-	Medium tolerant
Pollination	-	Wind
Country of Origin	-	Brazil (South America)
Area (India)	-	1062.04 Thousand Ha
Production (India)	-	817.00 Thousand MT
Productivity (India)	-	0.77 MT/ha
Top producer in India	-	Maharashtra> Andhra Pradesh> Odisha> Karnataka> Kerala
Top Productivity in India	-	Maharashtra (1.41 MT/ha)> West Bengal (1.14 MT/Ha)> Assam (1.08 MT/Ha)> Nagaland (1.08 MT/Ha)> Kerala (0.95 MT/Ha)
Top Producing in World	-	Philippines> Peru> Mali> Malaysia> Colombia
Top Productivity in World	-	Peru (82.41 MT/ha)
Propagation method	-	Softwood grafting
Spacing (m2)	-	7.5 m × 7.5 m
Planting time	-	July and August

Optimum temperature for growth	- 22-26°C
Harvesting period	- February – May
Yield (MT/ha)	- 10-15 kg per tree (2-3 tonnes/ha)
Storage temperature	- 67 % of RH at 30 °C
Storage life	- 6 months in the refrigerator
Inflorescence	- Panicle / polygamomonoecious
Edible part	- Cotyledon / Fleshy peduncle.
Fruit bud	- Mixed bud
Heterostyly	- Cashew

- Cashew kernel contains 47% fat, 21% protein and 22% carbohydrates.
- Cashew kernel does not have cholesterol.
- Collection of fallen nuts is known as Gllining.
- India is the largest producer as well exporter of raw cashew nut.
- The USA is the largest importer of cashew nut kernels in the world.
- Cashew is restricted to altitude below 700 m where the temperature does not fall below 20°C for prolong period.
- pH more than 8.0 is not suitable for its commercial cultivation.
- Pruning period: August-September
- Steam method is the most popular method of roasting.
- Best quality whole kernels kernels are obtained from drum roasting.
- Maximum recovery of oil is done using oil bath roasting.
- Inflorescence: Polygamomonocious
- Cashew nut is the III most important agriculture commodity exported from India.
- Broma dryer is a commonly used drier for drying of the kernel.
- The moisture content of the dried kernelranges between 2% to 4%.
- If the moisture content of dried kernel exceeds 75%, the kernel becomes susceptible to microbial attack.
- There are 26 grades of export quality cashewkernel.
- India's contribution to the global cashew exports is 65 % approximately.
- The dwarf rootstock of cashew – *Anacardium pumilum.*

- Cashew trees are very sensitive to waterlogging.
- Best production is seen in regions having altitude of 400m with at least 9 hr sunlight per day during the period spanning from December to May.India is the secondlargest consumer of cashew kernels in the world.
- All India Coordinated Cashew and Spices Improvement Project was founded in 1971 with its HQ at CPCRI, Kasargod, Kerala.
- In India, there are twenty-nine high yielding cashew varieties across various regions with mean yield ranging from 8-10 kg/tree/year.
- More than one million grafts of cashew are being produced every year using Standardized softwood grafting and flush/epicotyls grafting techniques
- The top working technique has been developed to rejuvenate the unproductive old gardens as well as poor yielders and plantations raised with seedlings.
- Control of the mosquito (*Helopeltis antonii*), causing 15-30% damage can be effectively controlled by the spray of endosulfan 0.05%.
- India is the world's leading producer, importer of raw and exporter of processed cashew accounting for nearly 50% of the world's demand and47% of the global consumption.
- In India, cashew processing factories alone employ 3 lakh people besides 2 lakh farmers that are employed in cashew cultivation.
- Mixed flower bud bears one-year-old shoot / last season growth.
- Ullal-1 is the first cashew variety released by any state for commercial cultivation.
- Cashew nut Research Station is located at Bapatala (A.P.).
- The fruit is a drupaceous nut; cashew apple is fleshy peduncle.
- About 80% of the fat contains unsaturated fatty acids. It is a good nerve tonic, steady stimulant and effective in controlling diabetes.
- The USA is the largest importer of cashew kernels in the world. 60% of the USA's cashew demand is met by India.
- Cashew nut grades include white wholes, scorched wholes, dessert wholes, whites pieces, scorched pieces and dessert pieces.

27

Persimmon

Common Name	-	National fruit of Japan / Ebony tree.
Botanical Name	-	*Diospyros kaki*
Family	-	Ebenaceae
Origin	-	China
Chromosome No.	-	90
Type of fruit	-	Berry
Climate	-	Subtropical
Inflorescence	-	Solitary
Respiratory	-	Climacteric
Edible part	-	Mesocarp and Epicarp.
Ploidy levels	-	Autohexaploid

- Deciduous, Monoecious fruit crop.
- Rich in Vitamin A (2710/U/100g)
- Fruits are highly astringent due to tannin content.
- Flowers appear in spring on new growth in February- March.
- Type of flowers: Male, Female, Bisexual.
- Pollination - Cross-pollination
- Mode of pollination - Insects
- Persimmon fruits exhibit a double sigmoid growth area.
- Date-plum (*Diospyros lotus*) used Rootstock in India
- American - *Diospyros virginiana*
- Japanese or oriental Persimmon - *Diospyros kaki*.
- Fruiting time – August- September
- Inarching on its rootstock on *D. lotus.*
- Dwarf and Semi-dwarf, modified central leader system.

- Astringent Cultivars: Hachia, Nightingale, Eureka, Hyakuma, Flat Seedless, Triumph, Honan Red, Saijo.
- Non-astringent cultivars: Fuyu, Wase Fuyu, Gosho, Hanafiah, 20th century, Jiro, Suruga.
- Pollination variant cultivars: Chocolate, Gailey, Maru, Hyakume.
- Sujika: Formation of white lines on the rind of the fruits.
- Most preferred fruit: PCNA type cultivars Fuyu & Jiro.
- Hachiya, (California) is commercially grown in India.
- 'Cincturing' is the removal of the ship of bark from around the trunk of a tree.
- Removal of astringency in persimmon fruits: Ethephon @500 ppm in 2 minutes.
- Alternate bearing is a common problem in which crop.
- Calyx cavity dehiscence is a serious problem.
- Skin rusting is caused due to high RH.
- Brix level at the time of maturity - 14-17°Brix.
- Temperature range of 8-11°C for 888 hrs in enough to break dormancy in persimmon.
- The genus Diospyros contains approximately 400 spp.
- In India, the European settler introduced persimmon in the hill states such as Himachal Pradesh, Jammu and Kashmir in 1921.
- The ripe fruit is delicious; the flesh is sweet and jelly-like. The entire fruit is edible except the seed and the calyx.
- China and Japan contribute a major share to the global persimmon market.
- Regions situated at 1000 to 1500m AMSL that have moderate winter which is favorable for Persimmon cultivation.
- It can tolerate temperatures as low as-15°C during dormancy.
- The buds are damaged by the spring frost.
- Non-astringent cultivars require warmer condition for fruit maturation than the stringent types of cultivars.
- A pH range of 6.5-7.5 is favourable for its growth.
- The *D. virginiana* is used as rootstock in Israel and USA.
- The seeds are extracted from fully ripened fruits during late October.

- The stratification of seeds is done over a period of 60-90 days.
- Tongue grafting is also done with a success rate of 60-65%.
- In India, planting is done in the months of January and February.
- Planting distance - 5.5 m × 6.0 m
- Dwarfing cultivars - Jiro-(5.0 m × 2.5 m)
- Semi-dwarf cultivar - Fuyu-(5.0 m × 3.0 m)
- Persimmon does not require high fertilizer doses.
- Application of balanced fertilizer in the ratio of 10:10:10 NPK is desirable.
- The fruits are clipped from the trees with shears leaving the calyx attached to the fruit together with a short stem.
- Great care is needed to avoid bruising.
- Persimmon fruits mature in mid-September.
- The tree continues bearing fruits for 4-5 years.
- Dwarf and semi-dwarf varieties bear for 2-3 years.
- A mature tree of Fuyu can yield upto 50kg fruit/tree.
- Jiro yields 80kg /tree.
- Hachiya yields upto 100kg /tree.
- Can be stored for 2-3 months at 30-32°C at 85-90% RH

28

Strawberry

Botanical Name	-	*Fragaria × ananassa*
Family	-	Rosaceae
Origin	-	France
Chromosome No.	-	56 (8X)
Type of fruit	-	Etaerio of drupe/ Etaerio of achenes
Climate	-	Temperate
Fruit is borne on herbaceous	-	Prostate perennial
Photoperiodic	-	Short day plant (SDP)
Growth patter	-	Single sigmoid growth curve.
Acid tolerant	-	Highly tolerant of acid soil.
Salt tolerance	-	Highly sensitive
Respiratory	-	Climatic fruit
Respiration rate	-	High – 20-40 mg of Co2/kg/hr.
Strong life	-	Very perishable – 0-4 weeks
Plant tissue (Fruit flesh)	-	Receptacle
Edible part	-	Fleshy thalamus
The aroma is due to	-	Ethyl butyrate
Ploidy level	-	Allo-octoploid
Genome sequences	-	250 Mbps. – 2011
Propagations	-	Runners

- *Fragaria chiloensis* × *Fragaria virginiana*
  (Diploid species) is the artificial man – made hybrid.
- Monoecious is a quick growing fruit plant and takes short span of the day to grow.
- It is a low growing perennial herb.

- It is one of the best fruit crops for the kitchen garden.
- Strawberry is a complete fruit with 98% of edible portions.
- It is rich in vitamin-C and iron.
- Ethyl ester *i.e.,* ethyl butanoate and ethyl hexanoate are responsible for the flavour of strawberry fruit:
- Indian wild strawberry *Fragaria vasca.*
- Hermaphroditism is much more common in *Fragaria chiloensis*
- Micro propagation was first started in this fruit.
- The flowers, generally white, are borne in small clusters.
- Both self and cross - pollination occur in this fruit crop.
- Honeybees are the major pollinators.
- Mulching is an important cultural operation in strawberry cultivation.
- A matted row system of training is commonly followed in India.
- China is the highest producer of strawberry in the world.
- Varieties: Chandler, Tioga, Torrey Selva, Belrlebi, Ferm, Pajaro Premier, Red Coat, Dilpasand, Local Jeolikot Bangalore, Florida-90, Katrain Sweet, Blackmore, Pusa Early Dwarf, Phenomenal, Majestic, Sujatha, Labella.
- Pajaro is the most successful variety under the summer system and has tolerance to viruses.
- Chandler is resistant to viruses, physical damages caused by rain and suitable for fresh market and processing.

  Arking (Cardinal × Ark-5431):

  Suitable for the meridional condition: **Tioga**

  Suitable varieties for septentrional culture: Senga Sengan, Red Gauntlet and Gorilla

  Day Neutral Varieties: Selia, Fern, Muir, Hecker, Tristar, Trileute

  Day-neutral and off-season variety: **Selesa**

  Ideal variety for processing: Midway, Midland, Hood, Redchief and Beauty.

  For the ice cream: Olympus Hood, Shuksan (High flavours and bright red colours)
- Strawberries are one of the most important soft fruits cultivated in the world.
- It grows well in different climatic zones ranging from temperate to tropical.

- Strawberry is a native of temperate regions, but some cultivars can be grown in subtropical areas.
- The cultivated strawberry (*F.* × *ananassa*) was first tried in Europe in the early 18th century and represents the accidental cross of *F. virginiana* from Eastern North America which, was noted for its fine flavour and *F. chiloensis* from Chile noted for its large size.
- Strawberry industries were setup at the beginning of the 18th century in America.
- Strawberry is a perennial herb, petiole is mostly long and channeled above, stipules adnate of the base of the petiole.
- Lewis 3-foliate, leaf fats sharply donate, but entire at the more or less wedge-shaped base.
- Strawberry requires minimum amount of dark period (8 hours) for flower bud initiation.
- 22-30°C temperature is considered optimum for plant growth.
- Blossoms of most June bearing cvs. are damaged at -2°C.
- The optimum temperature for the period between the onset of flowering and the onset of ripening should be 14-16°C.
- Strawberry thrives best in light soil as heavy soil inhibits root growth and development.
- Sandy loam soils rich in organic matter (4-5%) is best for strawberry cultivation.
- Strawberry is not very sensitive to soil reactions. Yet, it prefers a light acidic soils pH (5.5 to 6.4).
- For higher summer production, the inflorescence must be removed by hand to prevent fruiting.
- Application of GA3 at 50-100 ppm and not 60-120 ppm increases summer production.
- The roots of strawberries are confined in the top 40 cm of the soil, therefore, the 40cm soil surface should be fertile and friable.
- Strawberry is planted at a spacing of 25-30 cm plant to plant and 45-50 cm apart in rows from September to mid of October.
- Apply 50-60 tones/ha well rotter FYM 40 kg-P, 30 kg-K, 80 kg-N/ha.
- Foliar application of urea - (0.5-1.0%) and micronutrient formulation (0.3%) should be applied 3-4 times before flowering, fruit maturation and fruit enlargement.

- After one month of planting, the plants are mulched with black polythene or grass mulch.
- The ever-bearing cultivars (Ozark Beauty, Geneva, Super-faction) initiate and produce flowers through the growing season.
- Most cultivars produce hermaphrodite flowers. However, imperfect, pistillate and strictly female cultivars require cross-pollination.
- Incomplete pollination results in the development of irregular and poorly shaped berries.
- Placement of 2 honeybees hives /ha is recommended for effective pollination and a good fruit set.

29

Walnut

Common Name	-	King of Nut
Botanical Name	-	*Juglans regia*
Family	-	Juglandaceae
Chromosome No.	-	32
Fruit type	-	Nut
Origin	-	Central Asia
Climate	-	Temperate
Flower bearing	-	Both terminal and lateral
Inflorescence	-	Racemose – Catkins
Dichogamy	-	Protandry
Pollination	-	Cross-pollination
Pollination	-	By wind
Fat	-	64.54%, Protein – 14-20%
Max. Production	-	J&K
Harvesting period	-	Aug-Sept
Propagation	-	Tongue grafting

- Monoecious, temperate nut fruit.
- Most valuable exchange earning nut crop.
- The USA is the leading producer of walnut
- Female are born terminally and male flowers laterally
- Paradoxes is a rootstcck derived from *Juglans hindsii* × *Juglans nigra*.
- Reduction of walnut hull dehiscence Ethephon @ 2000 ppm.
- The blank nut is due to hot summers with low humidity.
- Trees deciduous 10-40 meter tall
- The temperature of 29-32°C near harvesting results in the wall filled kernels

- Sensitive to low temperature during spring and high temperature during summer.
- The chilling requirement is about 200-800hrs.
- Modified central leader system is ideal for training.
- Walnuts have high antioxidant content (Vitamin and fiber).
- Walnut contains omega 3 fatty acid (good fats).
- Immature fruits are rich in ascorbic acid
- It is believed to have originated in Iran and the areas surrounding it.
- The returning army of Alexander brought it to Europe from Iran.
- Almost all plant parts of walnut are utilized in one way or the other.
- Wood is used for making valuable furniture
- Walnut is grown successfully in all parts of the Himalayan region between the elevation of 1200-2150.

Climatic limitations for walnut are

a) Spring and fall frosts

b) Extreme summer heat

c) In-sufficient winter chilling.

  - The plants can tolerate as low as -11°C during deep dormancy.
  - The temperature of even 2 or 3°C below freezing point (0°C) kills leaves, shoots and flowers, thus resulting in crop failure.
  - The soil pH should be 5.5 to 6.5.
  - Exotic varieties: Hartley, Payne, Franquette, Serr, Sanland, Chico, Vina Howard, Chandler, Chana, Tulari, Lara.
  - Local selection: Gobind, Pratap selection, Solding selection, Katkhri selection, Kashmir budded, Wilson Wonder, Chakrata selection, Sulaman, Hemdam and Sirmour Selection.
  - The seeds are stratified for 90-110 days at low temperature of 4-5°C to break the seed dormancy.
  - Soaking of seed in 750-100ppm GA3 or 1000 ppm ethereal solution for 24 hours after stratification is very effective for stimulating seed germinations.
  - Side veneer grafting gives best results when performed during July-August.

- Patch budding performed at June- end to mid-July gives 80-85% bud take success.
- Grafted plants – 10.0 m × 12.0 m Seeding – 15.0 m × 15.0 m Paradox – 8.0 m × 8.0 m
- Fertilized with ½ kg of 15:15:15 NPK fertilizer mixture/year/tree up to 16 years of age.
- Fertilized with 8 kg of NPK mixture along with 100 kg FYM.
- The full dose of manure and fertilizer is applied during December-January.
- Pollen of *Juglans regia* is capable of fertilizing the eggs of the same variety and by other varieties as well.
- In walnut, dichogamy is the main problem responsible for the short period of both pollen viability and stigma receptivity.
- Improvement in pollination and fruit set can be made by

i) Hanging of catkins of male flowers.

ii) Planting of 3-4 varieties in the one hand.

- When about 80% of the hulls have cracked from the nuts, it is time for walnut picking.
- The spray of Ethephon 200-500 ppm on kernel maturity advancing harvesting by 10 days, makes largest of the entire crop at a time and promotes hull splitting.
- Application of nuts in Ethephon at 1500 ppm when the hull is still on causes hull to drop in 3-4 days.
- After cleaning, the nuts are dried with the help of driers at 32-30°C till the moisture content of nuts is about 8%.

30

Fig

Common Name	-	Badam
Botanical Name	-	*Ficus carica* L
Family	-	Moraceae
Origin	-	West Asia
Type of Fruit	-	Syconus
Edible Portion	-	Fleshy receptacle
Chromosome No.	-	X=19 2n=39
Per 100g		
Pulp	-	84%
Protein	-	4g
Vit. A	-	10010
Skin	-	16%
Calcium	-	20mg
Thiamin	-	0.10mg
Iron	-	4mg
T.SS	-	17.7%

Fig is divided by of 4th part

Edible fig (Common fig) Long style pistillate flower	**Capri fig (Wild fig) Short style pistillate flower**	**Smyrna fig Long style pistillate flower**	**San Pedro fig -**
1. Poona fig	1. Samson	1. Calimyrna	1. Dauphine
2. Canada	2. Stanford	2. Zidi	2. Lampeiria
3. Mission	3. Brawley	3. Taranimt	3. King gentile
4. Kadota			4. San Pedro
5. Brown turkey			

- Poona fig is the most popular cultivar grown in India (5.0 m × 5.0 m)
- Excel and Candia are suitable for HDP (2.5 m × 2.5 m)
- Candia is suitable for drying purpose.
- Excel is suitable for canning purpose.
- Deanna is suitable for table purpose.
- Fig is one of the first fruit to be preserved by drying.
- Salt and drought-resistant crop.
- Ficus carica is gynodioecious spp.
- Caprifig is monoecious spp.
- Common fig is pistillate.
- Excessive irrigation or heavy rains during ripening result in fruit cracking and producing of insipid fruits.
- Cultivar – Merselies, Black Ischia, Kabul, Banglara, Lucknow, (Dinkar – an improvement over Daulatabad).
- The fig is a subtropical fruit.
- Fig shows good growth when temperatures are above 15.5-21°C.
- Commercial propagation by Hardwood cutting, the use of 20-30 cm long, 0.5-7cm thick taken from 1- to 2-year-old.
- The cutting use of late summer treated by IBA.
- Planting time in North Western India is Monsoon season.
- Planting time for South India is August-September.
- Pruning in U.P. is performed in December, in Karnataka in Jan-Feb.
- Fig is fairly drought resistant and seldom irrigated many fig growing centuries.
- Four fruit growth stages during June, July, August and September by weight into 6 groups: 0-5, 6-10, 11-15, 16-20, 21-29, 30 and >30g.
- *Ficus glomerata* is a rootstock resistant to Nematodes.
- The major parts are stem-boring beetles (*Batocera rufomaculata* and *Batocera rubus*).
- Fungal disease rust caused by *Cerotelium fici* is a serious disease of fig in all the fig-growing countries.
- Sunburn is a serious problem in newly planted young trees. Source of infection is fungi; heavy pruning exposing the trunk and branches is responsible for sunburn. (Excessive exposure to).
- Fruit splitting or cracking is caused due to nutritional deficiency.

- The harvesting period for fig in North India is May, in South India- July-Sept & Feb-March.
- The average yield of figs is 180-360 fruits/tree; a harvest of 12 tonnes/ha.
- Storage for 4 weeks at 0°C and 90-95° RH with CO2: O2 ratio of 3.3 and 5.5.
- *F. glomerate* is resistant to nematodes.

31

Anola

Common Name	-	Badam
Botanical Name	-	*Emblica officinalis*
Family	-	Euphorbiaceae
Origin	-	Tropical Asia (Indo-China)
Type of fruit	-	Berry
Edible portion	-	Mesocarp and Endocarp
Chromosome No.	-	X = 7 2n = 28

- The aonla (Emblica officinalis) is an important minor fruit.

Moisture	-	81.2%	Phosphorus	-	0.02%
Protein	-	0.5%	Iron	-	1.2%
Fat	-	0.1%	Calorific value	-	59 mg/100g
Mineral matter	-	0.7%	Vitamin B	-	30 mg
Fiber	-	3.4%	Nicotinic acid	-	0.2 mg
Carbohydrate rates	-	14.0%	Vitamin C	-	600 mg/100g
Calcium	-	0.05%			

- The fruits are made into preserve (murabba) sauce, candy, dried chips, tablets, jellies, pickles, trophies powder etc.
- Banarsi, Chakaiya, Francis (susceptible to fruit necrosis) NA-4 (Krishna), NA-6 (NA-5 Kanchan), NA-9 (Neelum), N-7 (Amrit) free necrosis.
- Anola can be grown in the light as well as heavy soils except for very sandy one. However, well-drained fertile loamy soil is the best for its growth.
- Though Aonla is a subtropical fruit, it can be cultivated in a tropical climate quite successfully.
- Descending order of area wise production: Azamgarh > Pratapgarh > Varanasi > Bareilly
- Commercially propagated by Patch/T Budding

- Planting time is during July with a distance of 8m × 8m.
- The plant should be trained to a modified Leader system.
- By NDU in UP is the application of 10kg FYM, 125gN 50g P2O5 and 50g K2O per one-year-old plant.
- Age of 8 years (1000gN, 400g P2O5 and 400g2K2/O tree)
- Use of fertilizer is required during April-May and September-October.
- Flower bud differentiation in the 1st week of March.
- Sir ration (male to female) (1:307-1:197)
- Aonla is wind-pollinated crop.
- The fruit is capsular (drupaceous) with two seeds in each locule.
- The seed pulp ratio is 1:21.
- Since, there is no seed dormancy, fresh seeds give almost 100% germination.
- The productive life of anola trees is estimated to be 50 to 60 years.
- Aonla is tolerant to sodicity (35 ESP) and salinity (2 EC).
- Female flowers have tiny green perianth and the number of perianth segments varies from 5-7 with an average of 6.
- The ratio of female to male flowers on a tree may vary from 1:109 to 1:501 in different cultivars.
- Aonla is cross-pollinated by wind honey bee warp.
- Ber (Umran) has the highest number of hermaphrodite flowers (22.2%).
- It is grouped as an evergreen plant in the tropics but in subtropics, it exhibits deciduous nature.
- There is no self-incompatibility in aonla. The cause of poor fruit set may be attributed to a high percentage of staminate flowers.
- Flower and fruit drop
    - i) 70% of the flowers drop off within three weeks of flowering due to degeneration of the egg apparatus and lack of pollination.
    - ii) The second drop occurs from June to September due to lack of pollination and fertilization.
    - iii) Various stages beginning from the third week of August until October probably due to embryological and physiological factors.
- Fruit volume increased up to 75 days after fruit set.
- Fruit colours changed from green to yellow at 120th day.
- TSS increased upto 120 days of fruit set.

- Short Barer/Shoot gall maker (*Betousa stylophora*) control by – Rogor @0.03%.
- Rig Rust (*Ravenelia emblica*) a Saharanpur, India Dithane Z-78 at 0.2 per cent, Benznidazole @ 0.1% thrice at 15 days.
- Fruit necrosis a disorder associated with boron deficiency. It is controlled with three sprays of borax at 0.6% forth nightly.
- The budded plants start fruiting early *i.e.*, after 4 or 5 years.
- 10-year-old plant attains full mature bearing yield of 200kg per tree.
- Trees remain dormant from April-June and resume growth from last week of July to the first week of August.
- Aonla is drupaceous fruit (Capsule)
- Aonla trees flower twice a year during February-March in North and June-July in South India.
- Bearing – 3rd year after planting.
- In aonla, ber, guava and lemon are ideal filler plants.
- Aonla yield is 15-20t/ha
- UP has the maximum production and Area under cultivation.

32

Pomegranate

Common Name	-	Anar
Botanical Name	-	*Punica granatum*
Family	-	Punicaceae
Origin	-	Southeast Europe to Himalayas (Iran)
Fruit type	-	Balausta
Edible portion	-	Juicy seed coat
Chromosome No.	-	16, 18 X-8, 9
TSS	-	12.00% Juice
Moisture	-	85.4%
Total Sugar	-	10.67%
Pectin	-	1.4%
Titrable acidity	-	0.1%
Pectin	-	6.0% Seed
Reducing Sugar	-	4.7%
Protein	-	13.2%

- 52% use of the fruit of total weight.
- Comprising 78% juice and 22% seeds
- CIAH – Central Institutes for Arid Horticulture, Bikaner.
- Cultivars – Alandi or Vadki, Bassein Seedless, Dholka Ganesh, Jalore Seedless, Jyothi, Kabul, Kandhari, Paper Shell, Muskat Red, G-132, G-107, G-137, P-13, P-16, P-23, P-26.
- Hybrid – Mridul – Ganesh × Gul-E-Shah Red (fruit spot infection)
- Ruby – M
- Type of inflorescence is Hypanthodium.
- Pomegranate is a subtropical fruit crop.
- It is commercially propagated by hardwood cutting.

- Fruit cracking happens due to baron deficiency and water imbalance.
- HDP - (5 m × 2 m) - 1000/ha
  (5 m × 5 m) - 400/ha
- Both self and cross-pollination have been recorded in pomegranate.
- The varietal improvement program was started in 1905 at Ganeshkhind Botanical Garden, Pune (1920).
- Two ornamental types are Japanese dwarf and double flower, giving red, yellow or white flowers.
- Pomegranate is sensitive to soil moisture.
- Ganesh is a selection from 'Alandi' developed by Dr Cheema at Pune.
- Pomegranate is well grown in deep loamy and alluvial soil.
- The optimum temperature for fruit development is 38°C.
- A large number of root treatments are done using 200ppm IAA.
- These bear fruit for 3-4 years.
- Winter during the dormant period.
- Covering fruits with butter paper bags protects fruit from damage by Anar butter fly. (*Virachola isocrates*)
- Flowering season falls in June-July (Mrig bahar) coinciding with the break of monsoon February-March (Ambia bahar) and September-October (Hasta bahar)
- Anthesis to five days after the completion of anthesis and pollen germination is over 90 per cent.
- Pomegranate is a cross-pollinated crop (often cross-pollinated).
- The fruits are ready for harvesting 5-7 month after the appearance of blossoms.
- It starts giving 20-25 fruits per tree 4th year onwards.
- It starts yielding 100-250 fruit per tree 10th year onwards.
- Pomegranate fruit is non-climacteric.
- The respiration rate decreases continuously between 25 and 125 days after flowering.
- Flowering and ripening of fruits happen between 25°-29°C.

33

Datepalm

Botanical Name	-	*Phoenix dactylifera*
Family	-	Arecaceae
Origin	-	West Asia and
Type of fruit	-	Drupe
Edible portion	-	Pericarp
Chromosome No.	-	36, X-18
Moisture	-	7.30-9.50%
Riboflavin	-	135-173 mg/100g
Protein	-	2.30-2.78%
Fat	-	0.32-0.51%
Crude fiber	-	1.82-2.56%
Iron	-	10.5%
Minerals	-	1.80-2.28%
Total Sugar	-	86.10-87.90%
Reducing Sugar	-	73.40-82.72%

- The date is a good source of crystalline sugar.
- It is a dioecious and monocotyledonous fruit crop.
- Dates can be grown in sandy loamy soils.
- Date palm needs 3300 units of heat. (A unit of heat means a degree above a daily mean of 18°C between flowering, fruit development and ripening period.

Cultivars

- Halawi of makes tolerant of rain. It is 28.5 to 42.2 per cent Dakastage Dang Zoo, Hayni Khadrawy, Shamran (Sayen) are tolerant to high humidity. Medjoal, Bastree, Zahidi – is rain and high humidity tolerant.

Khalas

- Deep sandy lean soil is considered best for date growing.
- Maximum water holding capacity and good drainage are desirable.
- Date palm can be propagated by seed and offshoots.
- Seed germination of 91-55% in 1-9 months, flowering – 4-10 years.
- Planting time is during February-March and August-September.
- Planting distance of 7.5 m × 7.5 m.
- 7-14 days during mid-summer and 20-30 winter and dry soil in 23 days.
- 12000 m3 water is required per hectors per year.
- 600g N 100g P and 700g K per palm has the greatest benefit and 50-60 FYM.
- Leaf removal of 5 years, 20 leaves produced in/year, per plant in 70-90 leaves.
- Date palm is the inflorescence of primordial and Cross-pollination.
- Fruit set 50-80 per cent.
- The most common disease is false smut or graphical leaf sport.
- 40-70 kg fruit/palm.
- Best results of storage at 6-7°C in 6 months June-August in Doka stage harvest.
- One kg full ripe fruit provides approximately 3150 calories.
- 10% of male plants are raised in the garden.

34

Almond

Common Name	-	Badam
Botanical Name	-	*Prunus communis*
Family	-	Rosaceae
Chromosome No.	-	16
Origin	-	Central Asia
Type of fruit	-	Drupe
Climate	-	Temperate
Tree type	-	Deciduous
Inflorescence	-	Racemose – solitary
Heterostyle	-	Thrum type
Self incompatibility	-	Gemetophytic
Pollination	-	Cross-pollination
Growth patter	-	Double sigmoid growth
Fat	-	58.9%
Fiber	-	0.23%
Protein	-	2.8%
Calorific value	-	655mg / 100g
Phosphorus	-	0.49%
Carbohydrates	-	10.5%
Aroma	-	Benzal dehyde
Propagation	-	T-budding / Tongue grafting
Spacing	-	6.0 m × 6.0 m
Largest producer	-	USA
Pollination	-	Insects
Almond	-	Nut
Chilling requirement	-	800 hrs
Edible portion	-	Kernal or Cotyledon

Rootstock - Behmi (*Prunus mira*)
Shield budding - July to August
Training system - Central modified leader

- Recommended no. of beehives for efficient pollination is 5-8 beehives.
- Maturity indices: Change the colour from green to yellowish with cracks.
- Dehulling term is related to Almond.
- Damage to blossom due to early spring frosts is the major constraint in almond cultivation.
- Almond can withstand low temperatures upto -2.2 to -3°C.
- The almond tree becomes productive after 5 years.
- Almond contains oil - 49%, Oleic acid – 62%, Linoleic acid – 24% and Palmitic acid – 6%.
- Almond is a rich source of Vitamin E
- Kernels after blanching, roasting, frying and salting taste very delicious and are in in great demand.
- Green almond kernels are also consumed in the milky stage.
- Almond is cultivated mainly in regions situated between 360 and 450 N latitude.
- Flowers are hermaphrodite with white or pink petals, 5 sepals, a single and unicarpel pistil which usually contain 2 ovules.
- All cultivars of almond are selfsterile.
- Few cultivars like IXL and Non-Pareil are cross sterile.
- Varieties of J&K – Makhdoom, Parbat, Waris, Shalimar, Afghanistan seedling, IXL, Merced and Non-Pareil.
- HP – High and mid-hills – Merced, Non-Pareil, IXL valley areas – Drake, Katha, Peerless, Ne plus Ultra-dry temperate zone – Neplus-ultra, Texas, IXL.
- Most popular variety – Ne plus-Ultra
- Introduced – Makhdoom, Parbat, Waris, Shalimar, Phargaj.
- Hybridization – Sloh (Peach × Almond) – Self-fertile.
- Mutant V. – Supernova – Late flowering/self fertile.
- Self-fruitful – Drake, Kalha, Dhabar.
- Behimi – (Almond × Peach) natural hybrid.
- H98 – (Bruce × Sloh)

35

Pecannut

Common Name	-	Queen of Nuts
Botanical Name	-	*Carya illinoensis*
Family	-	Juglandaceae
Chromosome No.	-	32
Origin	-	North America
Fruit type	-	Nut
Climatic	-	Temperate
Tree fruits	-	Deciduous
Inflorescence	-	Catkins
Dichogamy	-	Heterodichogamy
Edible part	-	Cotyledon
Harvesting	-	October
Yield	-	20-25kg/tree
Propagation	-	Tongue grafting and Patch budding
Bearing habit	-	Alternate
Pollination	-	By wind
Pollination	-	Cross pollinations

- Male and Female flowers borne in a mixed flower bud.
- Most important nut fruit of world ranking: 5th in production worldwide.
- Chilling requirement – 400hrs at or below 7.2°C
- Contains 70% Fat and a good amount of phosphoric acid.
- HDP varieties: Desirable, Cheyenne
- Pecan is a valuable horticultural gift of North America to the world.
- Fat – 72%, Carbohydrates – 15%, Protein – 9%.
- Almost 90% of the nuts are sold shelled and rest in the shell.

- It is most commonly used in baking dishes and ice creams.
- Pecan shell, a byproduct, is also used to manufacture tannin, charcoal and abrasives in hand soap.
- pH ranges – 5-8
- Requires three coldest months between 7.2°C and 12.8°C with at least 400h of chilling.
- Varieties: Mahan, Nellis, Stuart, Western Schley, Mohawk, Cheyenne, Chickasaw, Desirable, Burkett, Wichita.
- Seedlings of various cultivars like Burkett, Nellis and Western Schley are used as rootstock for pecan.
- The bitter pecan (*C. aquatic*) is adapted to poor drainage, flooding conditions and low pH but gives low field.
- The seeds are stratified at 4°C for 70-90 days and soaking GA3 – 500ppm for 48 hrs.
- The pecan nut is planted in a flat land in the square system at 10-12m spacing.
- The pecan trees are trained in Central leader system.
- The wind is a pollination agent which carries pollens for about 900 meters.
- In addition, apply 500g NPK mixture (15:15:15) per year age of the tree up to 16 years.
- After 16 years, 8kg NPK fertilizer mixture should be added every year.
- Zinc deficiency – Zinc sulphate – 0.5%.

36

Chestnut

Botanical Name	-	*Castanea dentata*
Family	-	Fagaceae
Chromosome No.	-	24
Origin	-	Asia Minor
Type of fruit	-	Nut
Climate	-	Temperate fruit
Tree fruits	-	Deciduous
Inflorescence	-	Catkins
Dichogamy	-	Duodichogamy

- The chestnut is a nutritious low in fat and rich in Vitamin B.
- Freshly harvested nuts contain about 50% moisture, 40-42% carbohydrates, 2.9% proteins and about 1.5% fats.
- All cultivars of chestnut are self-sterile.
- The Chinese chestnut is the earliest tree to bloom. The flowers are produced in two kinds of catkins borne on current season shoot near the terminal portion of the shoot.
- Three nuts are usually produced in each bun (upper, left and right).
- Botanically, each nut is a complete fruit.
- The shell of the nut develops from the ovary wall.
- It is as hardy as peach and can withstand as low as -29°C temperature in deep dormancy.
- Some species of chestnut are used as rootstock for propagation.
- Chestnut is a highly cross-pollinated and hybrid seedling used as the rootstock is a possible cause of graft union failure in chestnut. Therefore, mixed hybrid strains should not be used as rootstock.
- The seeds are stratified for 50-60 days in moist sand at 0 to 2.2°C to break dormancy.

- The stratified seed is sown in nursery beds in March.
- The chestnut has only one bud at each node and bud break fails if it is killed by spring frost. So early grafting should be avoided.
- Chip budding also performs well.
- Chestnut is wind-pollinated as well as insect-pollinated.
- At least 1/kg 15:15:15 NPK mixture per year age of tree should be applied before sprouting or in early spring.
- The full bearing trees should be supplied with 100 kg FYM and 6.8 kg of NPK mixture during December-January.
- Chestnut is generally grown under rainfed condition but needs adequate moisture for at least 2 months after blooming.
- The chestnut matures in the first fortnight of October in Himachal Pradesh.
- Seedling tree yields 20-25kg of nuts at 12 years of age.
- Verities: Abundance, Crane, Kuling, Meiling, Nanking, Orrin, Colby and Hemming.

37

Hazelnut

Common Name	-	Filbert
Botanical Name	-	*Corylus avellana*
Family	-	Corylaceas / Betulaceae
Chromosome No.	-	22
Origin	-	Western Asia and Europe
Type of fruit	-	Nut
Climate	-	Temperate
Tree fruits	-	Deciduous

- Turkish hazelnut production of 625,000 tons accounts for approximately 75% of world productions.
- The common hazel (*Corylus avellana*) is native to Europe and Western Asia.
- In HP, it is found growing wild in Pangi region of Chamba district and locally known as Thangi.
- Hazelnut is also known as cobnut and filbert.
- Hazelnuts are extensively used in confectionery to make praline and also used in combination with chocolate truffles.
- The temperature of minus 10°C is critical, especially if accompanied by wind, which may kill both pistillate and staminate flavours.
- Hazelnuts are more shallow-rooted, most fruit and nut trees do not tolerate wet soils, the tree cannot tolerate excessive dry summer heat and hot winds.
- Hazel will grow in pH ranging from 4.5-8.5 but around pH 7 is ideal.
- Varieties: Tonda Romana, Barcelona, Negret, Tonda Giffoni, Tonda Gentile delle Langhe, Pavetet, Tombul.
- Planting – 6.0 m × 6.0 m and 5.0 m × 5.0 m.
- The fertilizer recommendations for hazelnut are 120 to 150 kg/ha of N, 60- 70 kg half and 100 kg/ha of K.

- Traditional training system – multistem bush.
- Best time of harvesting – mid-August.
- Five years of old plants produced average fruit field per plant 2.0 to 2.5 kg.

38

Pistachio Nut

Botanical Name	-	*Pistachio vera*
Family	-	Anacardiaceae
Origin	-	South East Asia
Chromosome No.	-	40
Type of fruit	-	Drupe
Climate	-	Temperate
Tree fruits	-	Deciduous
Cross-pollination	-	Dioecious
Dichogamy	-	Heterodichogamy
Edible part	-	Cotyledon

- It can withstand a very wide range of temperature from 30 to 42°C.
- Pistachio was grown for the first time in the early eighties using seeds of pista at the PCDO Bhoktu and Regional Horticultural Research Station, Sharbo in Kinnaur district of HP.
- The two commercial varieties namely Kerman (Pistillate) and Peters (Staminate) were introduced India from California in 2001.
- They are drought resistant and very tolerant of high summer temperature, but con not tolerant excess of dampness and humidity.
- *P. atlantica, P. terebinthus P. integerrima* are seedling rootstocks.
- Rafsanjani is the world-famous pistachio growing region in Iran.
- Iranian varieties: Momtaz, Owhadi, Agah, Safeed.
- Turkish varieties: Uzun, Kimizi.
- California varieties: Kerman, Peter Ibrahim, Owhadi, Safeed, Sharti, Wahedi, Bronte, Sfax.
- Other varieties: Buenzle, Minassian, Red Aleppo, Trabonella.
- A sister seedling of Kerman, namely Lasson also produces good quality large-sized nuts.

- The standard male cultivars are Peters. In the USA, 90% of the total pistachio cultivation area is under Kerman and Peters.
- Training – modified leader system.
- Bees in large number visit only the staminate flowers for the collection of pollen but are not attracted to pistillate flowers because they have neither petals and nor nectar.
- The wind is, therefore, the main cross-pollinating agent.
- One pollinating tree is planted for every eight pistillate (1:8) trees of this ratio may go up to 1:10.
- For synchronization of flowering, application of chemicals like Hydrogen Cyanamide and P^{333}
- The flowering can be delayed by spraying paclobutrazol at 100-4000mg/L during the end of June.
- Thinning out of 50% branches also fails to initiate more growth due to strong apical dominance.
- Application of a complete fertilizer such as 10-10-10 NPK should be adequate.
- P and K deficiencies result in unfruitfulness and poor field performance.
- The most serious disease of pistachio is the Verticillium wilt.

39

Jackfruit

Botanical Name - *Artocarpus heterophyllus* L.

Family - Moraceae

Origin - India

Type of fruit - Sorosis

Edible portion - Bracts / perianth / seeds

Chromosome No. - X=14 2n = 56

- Jack fruit a good source of pectin and contains about 1.9%.

		(Per 100g of edible portion)					
		Tender	**Ripe**	**Seed**			
Moisture	-	84	77.2	64.5	TSS	-	15-30%
Carbohydrate	-	94	18.9	25.8	Acidity	-	0.10-0.33%
Protein	-	2.6	1.7	6.6	Seed in starch	-	56.0-65%
Fiber	-	-	1.1	-			
Fat	-	0.3	0.1	0.4			
Iron	-	1.5(mg)	500(mg)	-			
Vit. C	11	-	-				
Vit. A	0	540	17				

- Jackfruit is eaten unripe at 25-50% full size as a vegetable or ripe as a fruit.
- The seeds are boiled or roasted and used in many culinary preparations. The seeds are very rich in carbohydrate and protein.
- Highest production in Assam.
- Jackfruit is a monoecious plant.
- The male inflorescence is long and light green and grows to a length of 5 to cm by 2.4cm.
- Pollen production is abundant and ranges from 1,50,00,000 to 1,70,00,000 per catkin. The percentage of pollen fertility ranges from 89 to 93 and the pollen diameter ranges from 16 to 22.
- Being cross-pollinated and mostly seed propagated, the jack fruit has innumerable types or forms considering the fruit characteristics.

- Gulabi, Champa, Hazari, Rudrakhs, Black Gold, Golden Nugget, J-31, Honey Gold and Lemon Gold.

Countries		Popular Varieties
Indonesia	-	Kandel, Mini, Tabouey
Malaysia	-	J-30, J-31, NS-1
Australia	-	Golden Nugget, Black Gold, Honey Gold, Lemon Gold.
Thailand	-	Dang Rashmi
Bangladesh	-	Topa, Hazari, Gala, Goal, Koa Khaja.

- This jackfruit can grow on a wide variety of soils although it prefers deep alluvial soil.
- Warm humid places are suitable for jackfruit.
- The largest area under jackfruit in India is in Assam *i.e.*, 8000 ha. Followed by Bihar – 4000 ha.
- Country wise area under jackfruit cultivation:
  Bangladesh→ India →Indonesia →Thailand→Philippines→Malaysia
- The most common method of propagation of jackfruit is by seed.
- Commonly, the square system is followed for planting spacing 12-12m of 70 plants/ha.
- The flowers generally start appearing in December and continue up to March and the fruit ripens in summer.
- Female spikes take 90 to 110 days to develop into mature fruit.
- Shoot and trunk borer can be controlled by arby's (sevin) 50%at the rate of 4g per liter of water by spray in flowering season is advisable.
- Fruit not caused by *Rhizopus artocarpi* is the most common disease of jackfruit in India.
- Normally, a tree bears a few to about 250 fruits annually at these stages.
- Storage life of about 6 weeks is expected when the temperature is 11.1-12.8°C and humidity between 85 and 90%.
- They cannot tolerate cold & frost.
- They are considered as "Good source of pectin".
- Seeds are sown immediately after extraction.
- Cauliflowers bearing habit.
- Soaking seeds in 25 ppm NAA for 24 hr improves their germination and seedling growth.
- The edible part of Jackfruit Bracts / Perianth.
- Yield/hac – 40-50 t/ha.
- This is a climacteric fruit.

40

Ber

Common Name	-	Badam
Botanical Name	-	*Ziziphus mauritiana*
Family	-	Rhamnaceae
Origin	-	Central Asia (Indo-China) (India)
Type of Fruit	-	Drupe
Edible portion	-	Pericarp
Chromosome No.	-	48 (4X) (Tetraploid)
Pulp	-	96-97%
Protein	-	2.9%
Moisture	-	81%
Vit. C	-	500-600 mg/100g
TSS	-	13-20%
Total sugar	-	3.3-5.8%
Non reducing sugar	-	3.3-8.4%
Ascorbic acid mg/100g	-	66-110

- In India, ber is cultivated on an estimated area of 61.284/ha.
- Fruit tree is suited for arid and semi-arid regions while crops in a tropical and subtropical climate.
- 125 cultivars grown in India: Nazuk, Kaithli, Umran, Chhuhana Ilaichi, Jogia, Banarasi Karaka, Najma, Sadhura Nannaul, Seo, Thornless, ZG3, Gola.
- The fruit starts ripening as early as the middle of December in the western part of India and North plain in mid-February.
- Propagation of Ring / T-budding.
- *Ziziphus nummularia* – Dwarfing due to formation Rootstock of an inverted bottleneck at the graft union.
- *Ziziphus rotundifolia* – Deep-rooted suited for arid zones.
- The seed germination of 3-4 weeks.

- In Punjab, the best time of planting ber is during February - March and in monsoon during July or September.
- Ber planting distance of 8 m × 8m.
- Best pruning time is during May.
- The dormant period is May-June.
- NPK – 480:150:300g/3year plant.
- Inflorescence in an auxiliary cyma.
- The flowering of Maharashtra – May-June.
- Other areas in flowering – September.
- The anthesis takes place between 7:30-8:30 am
- Flowering to printing takes 27 days.
- Fruit harvesting 150 days after.
- The fruit fly is the most serious part of September with the beginning of fruit set. 10% BHC @25kg/ha.
- Powdery mildew most serious disease the disease of attack in October-November. Sulphur Dust (250g/tree).
- Sooty mould or black leaf spot – (*Isariopsis indica*) captafol (0.2%) Carbendazim – (0.1%) Mancozeb (0.2%).
- 10 to 20-year-old trees of different ber cultivars vary between 80-200 kg/ tree.
- Storage temperature is 3.0°C for 1 month.
- Ber is also referred to as the king of arid fruit, poor man's fruit, Chinese fig, Chinese date.
- Antitranspirants (power oil 1.5% and Kaolin 7.5%) are used to reduce water loss and increase yield.
- Gala is tolerant against salinity and alkalinity and resistant to bank eating caterpillar.
- Incompatibility and Polyploidy are two serious hindrances in ber improvement.

41

Jamun

Common Name	-	Indian Blackberry, black plum, Java plum.
Botanical Name	-	*Syzygium cumini*, Skeel
Family	-	Myrtaceae
Origin	-	East India – Malaya
Type of fruit	-	Dupe
Edible portion	-	Epicarp and mesocarp
Chromosome No.	-	X-20, 2n-40
Moisture	-	84.5 - 86.4%
Protein	-	0.53 - 0.65%
Calorific value	-	83
TSS	-	9.0-11.5° (Brix)
To Sugar	-	5.8-6.9%
Acidity	-	2.1-2.5%
Reducing Sugar	-	2.3-3.7%
Vit. A	-	2.3-3.7%
Vit. C	-	73-100 (N), 30.3-40.7mg/100g
Calcium	-	0.02%
Iron	-	0.1%

- *S. fruticosum* (Jamiya), a small edible fruit, is grown as a wide break.
- Ross apple (Gulab Jamun) bears rose-scented fruits with fibrous flesh.
- Cultivar – Raij Jamun, Paras (Large size) N.J.-6 (Seedless).
- Syzygium Jambos – Rose apple (Gulab Jamun)
- Fruit trees for sodicity tolerance.
- Syzygium grows in tropical and subtropical climate under a wide range of them temperature and rainfall.

- Jamun, in general, prefers dry atmosphere at the time of flowering and fruiting.
- Water apple and rose apple are usually planted in home garden at a distance of 7.0 m × 8.0 m.
- Propagation of seed and shield and patch budding.
- Irrigation the use in May-June requires 5-6 irrigations in a year.
- Intercropping – peach, guava, kagzilime, as fillers, field crop – gram, peas, moong etc.
- Jamun exhibits maximum growth and flowering during February.
- Flower bud differentiation has been observed on 5-10 months old branches, starts in the last week of January and continues for 43 days.
- The flowers are hermaphrodite and light yellow.
- The maximum anthesis 18.71 - 43.08%, time 10.00am-12.00am.
- Jamun is a cross-pollinated and maxi fruit set 826-36%.
- Rajasthan®Gujarat®M.P.®U.P.®M.H.
- The fruit ripens in the month of June-July, seedling by 8-10 years and grafted by 4-5 years.
- Yield by seedling – 80-100kg by grafting – 60-70kg/year.
- Storage of 3 weeks at low temp. (9°$\pm$1°C and R.H. 85-90%).
- Being a good source of iron, it is used as an effective medicine against diabetes, heart and liver trouble.
- *S. densiflora, a* rootstock of Jamun is resistant to termite attack.
- Jamun is commonly grown in Rajasthan.
- Jamun (*Syzygium cumins* Skeels) is highly tolerant and suitable for wastelands.
- *S. fruticosum* a wild species with small but edible fruit is grown as a windbreak around orchards.
- *S. jambos,* commonly known as rose apple or gulab jamun produces fruits that are yellow in colour, inspired in taste with high pectin content.
- *S. jauanicum* (Water apple)
- The inflorescence in Jamun appears in the axils of leaves. The flowers are hermaphrodites.
- Jamun is cross-pollinated and pollination is accomplished by honey bees, hose flies wind and gravity.

42

Bael

Botanical Name	-	*Aegle marmelos*
Family	-	Rutaceae
Origin	-	India
Type of fruit	-	Amphisarca (Berry)
Edible portion	-	Succulent Placenta
Chromosome No.	-	X = 9, 2n = 39
Moisture	-	61.5%
Protein	-	1.8%
Fat	-	0.39 mg/100g
Minerals	-	1.7%
Carbohydrates	-	31.8%
Carotene	-	55 mg/100g
Thiamine	-	0.13 mg/100g
Riboflavin	-	1.19 mg/100g
Vitamin C	-	8 mg/100g
Niacin	-	1.1 mg/100g

- No other fruit has such a high content of Riboflavin.
- The Bael is a native of the Indo-Malayan region and has been known in India since prehistoric times.
- Narandra Bael 5– Size – 1.25 kg/fruit, very thin rind 2-5 mm, TSS-38%, Slight acidity-25 mg/100g, eight years old plant- 56 kg.
- Narendra Bael 9 - TSS-40%, Ascorbic acid – 28mg/100g.
- Pant Bael 12 – Varanasi, Gorakhpur, Mirzapur, Etawa, Kagzi Sewan Large, Deoria Large, NB1, NB5 and NB9.
- Bael is very hardy subtropical, deciduous tree and can thrive well even in swampy alkaline soils having pH range 5 to 10.

- Major bael growing states: U.P., Bihar, West Bengal and Orissa.
- Seeds, which are sown in June, usually propagate bael; Seedlings are transplanted a year later.
- Patch budding is the month of June or July gives the best results.
- Bael can be classified as a climacteric fruit.
- Bacterial shot-hale canker of Bael is caused by (*Xanthomonas blue*).
- 400 fruit/plant age of 15-20 years and 800 fruit – 40-50 years.
- Planting spacing – 8.0 m × 8.0 m.
- Leaves are used for offering to 'Lord Shiva'.
- Mature green fruits are ideal for harvesting.
- Storage temperature: 9°C + 85-90% R.H.

43

Karonda

Botanical Name	-	*Carissa carandas* L.
Family	-	Apocynaceae
Origin	-	India Java
Type of fruit	-	Berry
Edible portion	-	Epicarp and mesocarp
Chromosome No.	-	X-11, 2n-22.

		Fresh fruit	**Dry fruit**
Moisture	-	91.0%	10.2%
Protein	-	1.1%	2.3%
Carbohydrates	-	2.9%	67.1%
Fat	-	2.9%	9.6%
Iron	-	39.1 mg/100g	39.1 mg/100g
Calorific value	-	42 per 100g	364/100g

- The wine prepared from ripe fruits contains about 14.5 to 15% alcohol.
- Flowers are white, produced in a terminal cyme.
- It is grown in countries *viz.*, South Africa, Australia, India, Malaysia, S. Lanka and Barma.
- In India, it is grown in M.H., Rajasthan, U.P., Bihar, and South India.
- It is used as protective hedge in Gujarat and Punjab.
- It is not suited to the hills of Himalayas.
- Pant Sudarshan – White Pink
- Pant Suverna – Green, Purple
- Other varieties: PK.1, Pant Maohan
- Deep loamy or alluvial soils.
- It can tolerate saline-sodic soils with pH up to 8.5.
- Seeds usually lose their viability within 4-5 week.

- Seeds germinate 40-45 days after sowing, germination 68%.
- Transplanting is done after about two years from the sowing of seeds.
- Commercially propagated by seed and hardwood cutting.
- It is a non-traditional fruit.
- Planting distance of 1.0 m × 1.5 m and regular planting 1.5 m × 2.0 m.
- Flowering in Karonda starts from April-May and continues up to December under conditions similar to those of Chhota Nagpur (Bihar) conditions.
- The anthesis is between 12:30 to 4:30 pm.
- Only 36% of the flowers develop fruit to maturity.
- Carissa grandiflora – Natal plum.
- Carissa ovate – Jam Preparation.
- *Carissa edulis* – Scented flowers.
- Fruit has antiscorbutic property.
- Karonda behaves as a Xerophyte plant.
- Harvesting time is during Aug-Sep, 4-5 kg/plant.
- Cold storage 13°C and 94% R.H.

44

Mulberry

Botanical Name	-	*Morus alba* L.
Family	-	Moraceae
Origin	-	China
Type of fruit	-	Syncarpous/aggregated drupe/sorosis
Edible portion	-	Mesocarp
Chromosome No.	-	308
Moisture	-	85-88%
Carbonyl	-	7-8-9.2%
Protein	-	0.4-1.5%
Fat	-	0.4-0.5%
Acid	-	1.1-1.9%
Fiber	-	0.9-1.4%
Minerals	-	0.7-0.9%
Ripe Sugar	-	8-9%

- The most important use of mulberry leaf is rearing of silk worm for production of silk.
- The mulberry tree is used for landscaping in Asia, Europe & Southern U.S.A.
- Morus alba (White mulberry)
- Morus rubra (Red mulberry) is native to America.
- Morus nigra (Black mulberry)
- Mulberry plants are monoecious in sex expression.
- Inflorescence of catkins.
- Major mulberry producing countries; China®India®brazil
- The most extensive area under mulberry cultivation in India is in Karnataka.
- Major mulberry growing Indian states are: T.N.→Kerala → U.P.→ Bihar→ Madhya Pradesh→W.B.→Rajasthan.

- Cultivar→Kanva-2, It is grown in K.N, A and T.N. S-30 and S-54, S-1, W.B., S-799, Kosen (a Japanese variety) – W.B., K-2, M-5.
- Commercial propagation is done by stem cutting/budding.
- Planting spacing of 6-8m is appropriate.
- The root-knot nematode is caused by Meloidogyne incognita.
- Mulberry is a non-climacteric fruit which is consumed either fresh or after processing.
- A full-grown tree can produce 30-40kg fresh fruit/plant and leaf 250-350g/ha.

45

Phalsa

Common Name	-	Star Apple
Botanical Name	-	*Grewia subinaequalis*
Family	-	Tiliaceae
Origin	-	India
Type of fruit	-	Berry
Chromosome No.	-	36
Juice	-	50-60%
Sugar	-	10-11%
Acid	-	2.0-2.5%
Moisture	-	80.8%
Protein	-	1.3%
Carbonyl	-	14.7%
Vit. C	-	22 mg/100g
T. minerals	-	1.1%
P. Carotene	-	419 (μg)

- Ripe phalsa fruits are sub-acidic and a good source of Vitamin A & C.
- Kernel contains 2.0 %and 23.7% oil respectively which is rich in linoleum acid.
- Colour phalsa fruit is anthocyanin pigment.
- The phalsa seed contains 7.17 per cent crude fat with a typical.
- The plants can tolerate temperatures as high as 44°C since high temperature favours ripening of the fruits.
- It is a subtropical crop.
- It is commercially propagated by seed.
- The viability of seeds can be retained for 6 months under cool storage.

- Seed germination of 15-20 day get ready for planting in the field.
- Seed germination per cent 76%.
- Shortest germination period shortest of only 16 days.
- It completes flowering and fruiting within 4-5 month.
- Planting is usually done 2.5-3.0m, 1100-1500 plant/ha.
- Phalsa plant is normally pruned at 0.9 to 1.2m above the ground level; time of pruning is during December—January when they are dormant annually.
- Application of N, P and K at 100, 40 and 25kg/ha.
- Flowering in phalsa starts February - March onwards and continues till May.
- The fruits mature after 60 days.
- The fruit is non-climacteric.
- The only serious pest is bark eating caterpillar (*Indarbela* spp.).
- Brown spot of phalsa is caused by (*Cercospora grewia*) disease.
- Harvesting time is from April to fruit week of June. The fruits do not ripen at the same time.
- Yield of Single row system is 6.0 to 7.0 kg/plant and Double row system is 4 to 4 kg/plant.
- It is inflorescence of cyme.
- Two types, dwarf and tall phalsa are known to exist.
- It is very sensitive to Fe deficiency.
- Phalsa is a highly self-pollinated crop.
- It is an ideal plant for multi-storey cropping.

46

Tamarind

Common Name	-	Tamaria, Tamarindier, Tamairindo and Imli.
Botanical Name	-	*Tamarindus indica* L.
Family	-	Leguminosae
Origin	-	India/ Tropical Africa
Type of fruit	-	Pod
Edible portion	-	Pulp or Mesocarp
Chromosome No.	-	X-12 , N-24
Moisture	-	20.9%
Protein	-	3.1%
Carbohydrates	-	67.4%
Fiber	-	5.6%
Fat	-	0.1%
T. Minerals	-	2.9%
Calorie	-	230-283 cal/100g
Total acidity	-	12.2-23.8%
Com. Name	-	Tamaria, Tamarindier, Tamairindo and Imli.

- Tamarind is at the 6th position in terms of export value, fresh, dried, powder and paste.
- It is a good source of iron.
- The red variety of Tamarind fruit contains anthocyanin while leucoanthocyanidin is present in the brown.
- Leaves are very rich in anthoxanthin, while the fruits are poor in this pigment.
- Xanthophylls are responsible for the yellow colours of the flower.
- The pulp contains 1.27% pectin.
- T.K.P. – Tamarind kernal powder, the use of cattle freed also.
- Kernal

i) Protein - 17.1-20.1%,

ii) Fat - 6.0-7.4%,

iii) Carbohydrates - 65.1-72.2%.

iv) Crude fiber - 0.7-4.3%,

v) Ash - 2.5-3.2%.

- The fruit is 8.20cm long, brown irregularly caused pods when fully ripe.
- Cultivars – Parthisthan, Yogeshware, P.K.M.1 and Urigam.
- Tamil Nadu®A.P.®M.H.®K.N.®Kerala®Orissa®M.P.®Bihar
- Tamarind is the propagation of seed and softwood grafting ® In-situ softwood grafting.
- It should be planted at 12 m × 12 m.
- Tamarind of tanning is an essential practice in the initial 2-3 years.
- In Tamarind blossom bud differentiation begins with the resumption of vegetative growth just after leaf fall and flowers appear from April to July in Southern, Northern and Western India.
- Tamarind is a protogynous and highly cross-pollinated crop by – flies, bees and red nuts.
- Fruit set under natural conditions low ranging from – 3 to 5%.
- Alternate bearing.
- Lac insect (*Laccifer lacca*) is seldom a major part of Tamarind – Chlorpyrifos – 5ml/L.
- In India, fruits generally mature in F eb-Mar while in the tropical north and South America.
- The ripe pods are collected from the trees by shaking the branches.
- The ripened pods if not picked may remain hanging for almost one year after flowering.
- A full-grown tree produces about 200kg of fruit per annum.

47

Woodapple

Botanical Name	-	*Feronia limonia* L.
Family	-	Rutaceae
Origin	-	India
Type of fruit	-	Amphisarca (Berry)
Edible portion	-	Lobed cotyledons
Chromosome No.	-	X=16, 2n=32
Moisture	-	64.2%
Protein	-	7.1%
Carbohydrates	-	17.0%
Fat	-	0.3%
Total minerals	-	0.3%
Calcium	-	4mg
Phosphorus	-	9mg
Acidity	-	2.3%
Sugars	-	7.2%

- Its constitution makes the tree one of the useful medicinal plants of India.
- Soils including degraded soils of the arid region and can tolerate salinity to a certain extent.
- It is mostly grown in dry portions of subtropical and tropical regions of the country.
- The mature plants can tolerate temperatures as low as 0-15°C and as high as 47.5°C.
- Commercial propagation of seed / In-situ budding.
- Ideal Planting distance required is of 8 to 10m
- Opening of flowers starts in the 2nd week of March.

- The flowers are mainly staminate and hermaphrodite.
- Ripe fruits are available from October-March.
- The tree can produce about 200 to 250 fruits per annum.